CONSTRAINT-BASED VERIFICATION

CONSTRAINT-BASED VERIFICATION

JUN YUAN
Atrenta, Inc.

CARL PIXLEY
Synopsys, Inc.

ADNAN AZIZ
University of Texas at Austin

 Springer

Jun Yuan
Atrenta, Inc.
Suite 440 W
2001 Gateway Place
San José CA 95110

Carl Pixley
Synopsys, Inc.
2025 NW Cornelius Pass Rd.
Hillsboro OR 97124

Adnan Aziz
University of Texas
Dept. of ECE
ACE 6.120
Austin TX 78712

Constraint-Based Verification

e-ISBN 0-387-30784-2
ISBN 978-1-4419-3852-7   e-ISBN 978-0-378-30784-8

Printed on acid-free paper.

© 2006 Springer Science+Business Media, Inc.
Softcover reprint of the hardcover 1st edition 2006

9 8 7 6 5 4 3 2 1

springer.com

To my wife, Ting, and my daughter, Justine. To my parents. And to Katy.

. Jun

To my wife, Cynthia, and my children, Laura and Margaret, for many years of love, patience and support. Also to some wonderful teachers, Miriam Heisar, Prabir Roy and RH Bing, who taught me to love mathematics.

. Carl

To Professor Robert Brayton.

. A. A.

Contents

Chapter 1

INTRODUCTION

Electronic designs have been growing rapidly in both device count and functionality. This growth has been enabled by deep sub-micron fabrication technology, and fueled by expanding consumer electronics, communications, and computing markets. A major impact on the profitability of electronic designs is the increasing productivity gap. That is, what can be designed is lagging behind what the silicon is capable of delivering.

The main cause of this productivity gap is the cost of design verification. Verification complexity grows faster than the design complexity, which in turn grows exponentially, as Moore's Law has successfully predicted. This leads to the verification crisis, a phenomenon that has become ever so familiar in today's Electronic Design Automation (EDA) landscape.

There are several remedies, each coming from different aspects of the design and verification process. The first is the movement to higher levels of abstraction, especially the emerging Electronic System Level (ESL) model. The key enablers include languages that capture system level behavior and facilitate testbench automation for high level verification.

The second are the methodology changes, exemplified by assertion-based verification, and testbench automation highlighted by constrained random simulation. Both can find specialized constructs in, and are facilitated by, the ESL modeling languages.

The third is the advance of technology at the foundation of all the changes. Constrained random simulation, with robust constraint solving capability, is key to any practical testbench automation tool. The same fundamental solving techniques are also shared by formal verification tools in assertion-based verification. The formal semantics for assertions, now entrenched in the ESL languages, connect interface constraints used in constrained random simulation, and properties monitored in both simulation and formal verification.

1.1 Design Complexity

Integrated circuit design is continuing its rapid growth in scale and functionality. The transistor count per chip has been doubling every eighteen months for the past three decades; this pattern can well be extrapolated for years to come. System-on-Chip (SoC) designs today can have up to 200 million transistors on a single chip. A study from International Technology Roadmap predicts that the number will reach 1 billion by the end of the decade.

There are two driving forces that have brought Moore's prediction to fruition: the continued advance of integrated circuit fabrication technology and the consumerization of electronic products. The minimum feature size, specifically the length of a MOSFET, has been consistently shrinking, from over 1 micron in the early 1980's down to today's deep sub-micron territory of 65 nanometers and below. In the late 1990's, many predicted that the feature size could not go below 0.17 microns because this was the physical limitation of the lithographic manufacturing process, but this limitation has been overcome. We seem to have unlimited resources in miniaturizing electronics. The result? Room-sized computers have been reduced to desktop, laptop, and hand-held computers, cellular phones have shrunk to the size of a watch, and digital cameras have enough pixels and processing power to rival film-based professional cameras, all of which have fueled the rapid growth of consumer electronic products.

Electronics, over their history of production, have been getting smaller in size and faster in performance. The types of products being made have also evolved over time. During WWII, electronics were made for purely military use. Then business computing machines were manufactured, followed by calculators, personal computers, and a plethora of other consumer products. As can be seen in Figure 1.1, it is in the personal computer era in the 1990's and the consumer market today that electronic product sales have enjoyed an exponential growth.

However, unlike personal computers, whose performance is primarily judged by improvements in speed and increases in memory, today's consumer products have a more diverse agenda, as exemplified in the list below.

- Top-of-the-line computing power

    For example, 64-bit game boxes with state-of-the-art graphics processing capability.
- Rich functionality

    For example, the all-in-one cell phone with camera, PDA, email, and web browsing.
- Low power

    For mobile computing and cheaper cooling.
- Shorter life cycle

    Due to unpredictables in trends and fashion.

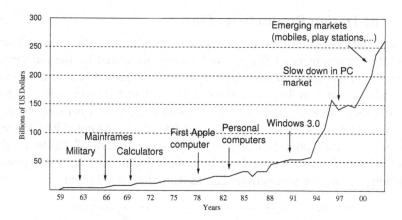

Figure 1.1. Revenue history of electronic systems.

The short duration of product generations and the constant demand for greater functionality has given rise to the so-called "System-on-Chip" movement. Simply put, it is not feasible to design every transistor from the ground up in today's multi-million-transistor chips. Instead, in-house or third-party Intellectual Properties (IPs) are reused, and designs are beginning to be done at the system and integration levels. The result is a complex system that may contain tens of IPs (Figure 1.2). This trend of IP reuse and integration will soon transform the SoC into a Network-on-Chip (NoC), where a single chip holds a "sea of processors" connected by networks.

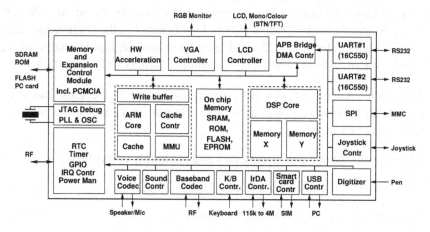

Figure 1.2. An SoC design with many IPs (Courtesy: Synopsys Inc.).

1.2   The Design Productivity Gap

Although the SoC-centric paradigm shift is transforming the chip industry, the yearly improvement of design productivity, measured by the number of transistors per designer month, is lagging behind the growth rate of transistors per chip by as much as 37%, as estimated by Sematech (Figure 1.3). Another study has found that by 2005, when 250 million transistors per chip become feasible, most SoCs would use only around 50 million. As noted by Chinnery and Keutzer in their book "Closing the Gap Between ASIC & Custom", this lack of utilization manifests itself as a loss of quality in speed, area, and power. Ultimately, this design productivity gap leads to unrealized profit.

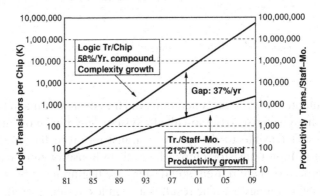

Figure 1.3.   The design productivity gap.

While the EDA society is primarily responsible for this unrealized profit, we still see an out-of-proportion investment on EDA tools by the semiconductor industry. As shown in the "reverse pyramid" in Figure 1.4, the net revenue seen by EDA vendors is less than 2% of chip revenue, and negligible compared to electronic system revenue. There may be multiple reasons for this phenomenon:

- The semiconductor industry is operating on low profit margins, and thus the EDA cost may represent a considerable portion of the semiconductor manufacturer's profit.

- EDA startups have a low startup cost and are highly competitive.

- EDA business and sales models are at fault.

- EDA vendors are not producing the right tools.

It is possible to see the current imbalance in EDA investment in a positive light: for designers and project managers, EDA tools are still cheap and affordable, while for EDA vendors, there is great potential for growth.

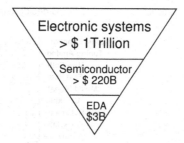

Figure 1.4. The reverse pyramid of revenues.

1.3   The Verification Crisis

Who then is responsible for the design productivity gap, and what is the remedy? We have pointed to a business environment of low EDA revenue, which, of course, has an impact on the vendors' ability to create and market their tools. On the other side, the growth of design productivity itself is facing a formidable challenge – the so called "verification crisis."

The verification crisis is not looming on the horizon, it is a reality that many in the electronic design community have been painfully living with for years. According to a Collett International study (Figure 1.5), the rate of first silicon success is steadily declining, dropping to 35% in 2003; 70% of re-spun designs contain functional bugs, yet by that point 70% of the total project effort for complex chip designs has already been spent on verification. Verification has become such a dominant factor that doubling the verification efficiency would by itself compensate for the 37% deficit in design productivity relative to the manufacturing productivity mentioned in the previous section.

Broadly speaking, the verification crises can be attributed to the following interacting situations:

- Verification complexity grows super-linearly with design size

- The increased use of software, which has intrinsically higher verification complexity

- Shortened time-to-market

- Higher cost of failure (low profit margin)

In the paradigm of functional verification, complexity is often measured by the size of the design space, which is exponential in the number of storage elements (flip-flops) in the design. If we assume that design behavior is typically modeled as sequences of states, the complexity becomes even higher. Since

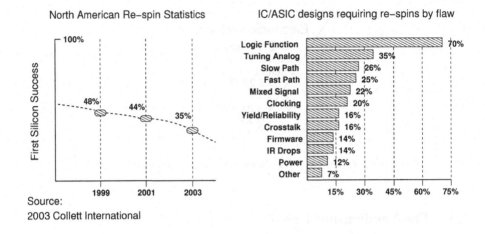

Source:
2003 Collett International

Figure 1.5.   Silicon re-spin and causes.

design size grows exponentially, theoretically we should expect a double exponential growth of verification complexity. However, in reality this worst case
scenario is often avoided due to the common practice of divide-and-conquer;
for example, the data path is usually verified separately from the control, and
blocks are verified individually, and the integration is verified at the end.

Still, empirical study has shown that doubling the design size doubles not
only the amount of verification effort per clock cycle but also the number of
cycles required to reach acceptable coverage [But89, SI98]. Roughly speaking,
this results in a verification complexity that is Moore's Law squared [MBC$^+$00].

The above estimate does not include the co-verification of embedded software, which plays an increasingly important role in SoC design. IP reuse also
brings with it a unique verification challenge: IPs that were once stand-alone
chips are now blocks in an SoC, and the integration verification is feasible only
at a modeling level that is more abstract than the register-transfer level.

Additionally, the design cycle constantly shrinks in response to the extrav-
agant demands of the consumer market, where features and performance may
differ considerably from one production generation to the next. As it stands
today, the product life cycle in many consumer markets is well below one year.

These short production life cycles aggravate the problems brought on by
design failures. A silicon re-spin can easily cost $1 million, and this is negligible
compared to the cost of missing the window of profitability. Aurangzeb Khan,
executive vice-president of Simplex's SoC Design Foundry stated in his 2002
Design Automation Conference presentation that a three month delay of a high-
value, high-volume product could cost as much as $500 million. If the cost is

prohibitively high, the whole project may be scrapped. There are also high profile examples of recalls that cost hundreds of millions of dollars.

Clearly, there is presently a formidable verification crisis. The solution must come from all possible factors in the design and verification space.

The most direct response would be to enhance the capacity of the underlying verification methods. Faster simulation, emulation, and static verification all fall into this category.

A second response would be to incorporate the methodology changes made possible by advances in technology. The most prominent examples are constrained random simulation in testbench automation, and assertion-based verification using both static (formal) and dynamic (simulation) methods. The third response would be to shift to a higher abstraction level for design and verification, as evident in the recent advance of behavior synthesis and transaction level modeling. Last but not least, the enhancement and creation of design and verification languages will play an indispensable role in facilitating all of the previously mentioned advances.

1.4 Design Modeling and Verification

Strictly speaking, a design model is an abstraction. At the very bottom of the abstraction hierarchy is the transistor level modeling of micro-electronic circuitry. Higher up the hierarchy are the layers of gate level, register-transfer level (RTL), and so-called behavioral modeling.

Verification occurs at each one of these abstraction levels. Examples include SPICE simulation with transistors, HDL simulation with gates and RTL, and C/C++ simulation using behavioral models. The evolution of EDA tools has been dominated by the trend of moving up this design and verification hierarchy of abstraction. Consequently, design entities have changed from transistors, to gates, to RTL descriptions, and to even higher level components, and verification time scales have increased from nanoseconds, to events, to cycles, and to transactions.

Of course, improved design productivity due to abstraction does not come for free. With each higher abstraction, the following challenges must be addressed:

- Define and standardize the new modeling: for example, SPICE, Verilog/VHDL, SystemVerilog, etc.

- Refine from the new abstraction to a lower level abstraction: for example, RTL and behavior synthesis.

- Verify the refinement, that is, verify that versions at different abstract levels correspond.

Figure 1.6 shows the hierarchy of abstraction levels and modeling methods, and the associated verification and synthesis tools.

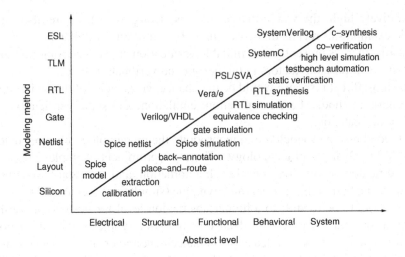

Figure 1.6.   Hierarchy of abstraction, modeling, and tools.

Design and verification are mainly done in RTL, which has been the case for over 10 years. RTL-based design, when it was first introduced, brought about a wave of productivity innovations. The most prominent examples are logic synthesis and optimization that beat hand-crafted designs in quality, while taking a fraction of the time, and RTL simulation that runs orders of magnitude faster than gate and SPICE simulation. However, RTL-based design has reached its limit, as the increasing productivity gap makes evident.

Behavioral modeling has been championed for years, and though design tools for it have been introduced, they have proven too premature to make a significant impact. The same three questions remain to be answered about behavioral modeling: What is it? How is it used to refine circuitry? How is it used to verify circuitry?

The emerging ESL design, being more abstract than RTL, seems to provide at least partial answers to these questions.

- There is a general consensus that an electronic system is hardware running embedded software, and that the observable activities are transactions or instructions. The designers of ESL modeling languages kept this idea in mind when drafting SystemC and SystemVerilog.

- Synthesis tools targeting C/C++, SystemC, and SystemVerilog are maturing; some can generate RTL designs that are as good as, or even better than, custom designs.

- All proposed ESL languages address the high level verification problem by providing assertions and testbench automation, highlighted with transaction-level verification and constrained random simulation.

ESL has taken a more conservative approach. The intelligence and technologies embodied in these answers are collective and have been accumulating over years. Still, the gradual change has resulted in a revolution that is rippling through the entire spectrum of EDA solutions.

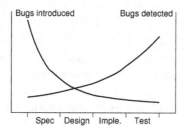

Figure 1.7. Dilemma of the sequential approach to verification.

That the ESL verification of system level requirements is more efficient than verification at the RT level is obvious. One fact that should be stressed is that ESL verification addresses a classic dilemma that conjoins all the verification constraints – the sequential approach to design and verification. On one hand, most functional bugs are introduced at the specification and early RTL stage. On the other, substantial verification effort commences only after the RTL has become stable and the physical design has begun (Figure 1.7). Consequently, a specification bug found at this late stage will force considerable debug and re-design efforts that will cycle through the architecture, RTL, synthesis, and timing stages. Verification at ESL is specifically aimed at rooting out these costly high level bugs.

What makes ESL most appealing is the idea that RTL sign-off to ASIC vendors will one day become a standard practice when the move to ESL is complete. Gary Smith from Gartner Dataquest has predicted that ESL, which offers a way to address the design complexity and verification crisis, is the next critical advance in electronic design automation.

1.5 Dynamic versus Static Verification

Functional verification approaches can be broadly classified as being either dynamic or static. In both cases, the verification challenge stems from the drive for completeness. In dynamic verification, that is, simulation, the exploration of complete design functionality, using case-by-case testing, translates indirectly into having to examine the design on all possible execution sequences with

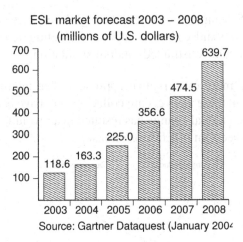

Figure 1.8.   ESL sales forecast (Gartner Dataquest, January 2004).

all combinations of the design inputs. This is a formidable task even for designs of modest size, as even a 256-bit Random Access Memory (RAM) yields approximately 10^{80} possible combinations of its contents.

Since exhaustive simulation is practically impossible to achieve, a sense of relative completeness is usually established through the use of some "coverage" metric. Defining a coverage metric is a selective, and thus subjective, process. Another major task in simulation is to develop a "testbench" for the generation of test stimuli. In itself a time-consuming process, testbench development is further complicated by the requirement of adequate coverage.

Because of the problems inherent in simulation, functional verification sometimes resorts to the complementary static approach. Static verification is often assisted by non-enumerative, analytical techniques, such as theorem proving and model checking. These methods are intrinsically complete with respect to the set of properties being checked, since they are founded on rigid mathematical analysis or on an exhaustive search of design space.

However, the completeness of static verification is also its main drawback. Its computational complexity is usually much higher than that of simulation, making static verification inadequate to the task of handling large designs. Consequently, static verification (with the exception of equivalence checking) is used most often in designs that are small or abstracted lower level blocks, such as bus arbitration and interface protocols.

Simulation remains the mainstay of verification, whereas static verification plays a complementary but increasingly important role. When combined, simulation and static verification can offer benefits not available to each independently.

1.6 Simulation

Simulation checks whether the design exhibits the proper behavior as elicited by a series of functional tests. More specifically, simulation consists of four major tasks:

- Generating the functional tests

- Executing the test stimulus on the design

- Determining if the design behavior satisfies its specification during the execution

- Collecting coverage statistics

The second task, execution, is performed by a simulator. The other three tasks are performed by a testbench.

1.6.1 Simulators

HDL simulators can be event-driven, cycle-based, or a hybrid of both. An event is a transition of a logic signal, which in turn may cause another event to happen later. An event-driven simulator both schedules and executes events. Due to the dynamic nature of scheduling, event-driven simulation can model delays, asynchronous inputs, and feedbacks.

The efficiency of event-driven simulation arises from the observation that, typically, only a low percentage of all signals change in a given clock cycle. There are, however, drawbacks. The scheduling requires constant data dependency checks, and many events are transient and don't affect the stable state of the design.

A cycle-based simulator samples signals only at specific points in time. Usually these points are the clock edges in synchronous designs. The sampling is performed at the output of every flip-flop and by considering all the logic in its fan-in cone, which is often done in a fast "compiled valuation," as opposed to the gate-by-gate interpretation of event-driven simulation.

Cycle-based simulation is aimed at the functional verification of synchronous designs, and does not handle precise timing. Consequently, it has difficulty in modeling asynchronous changes such as feedbacks and transparent latches.

State-of-the-art simulators are usually a mixture of event-driven and cycle-based simulation. The former is applied to the synchronous portion of the design, while the latter is used when asynchronous behavior must be simulated. To avoid transient events and data dependency analysis, signal evaluation is levelized as much as possible during pre-simulation. Other optimization techniques include native compilation and reduction from multiple-state to 2-state.

1.6.2   Testbench

Most electronic designs are reactive systems that respond passively to their environment. In verification, the environment is modeled by a testbench, which drives the inputs of the Design Under Verification (DUV). At the unit level, a DUV can be a processor, a Direct Memory Access (DMA), or other peripheral devices, where the environment is usually a bus. At the block-level, the environment consists of the neighboring blocks of the DUV.

Testbench writing is a refinement process starting with a specification of the design written in natural languages, for example, English. The features and functionalities to be tested are extracted from the specification that, together with the interface protocol, defines the relationship of the DUV with its environment. These functionalities are then grouped into test cases, and testbenches are written such that the interface protocol is respected and the test cases are exercised. The key ingredients of a testbench, as shown in Figure 1.9, are test generation, result checking, and coverage analysis.

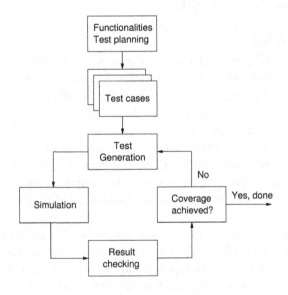

Figure 1.9.   A testbench flow.

1.6.3   Test Generation

Test generation methods can be divided into two broad classes: directed and random. Directed tests are developed manually in order to exercise specific scenarios corresponding directly to the functionalities being tested. These tests are typically deterministic and closely coupled with the DUV, resulting in poor reusability.

Random test generation compensates for the inflexibility of directed tests by randomizing the generation subjected to the satisfaction of the DUV's interface protocol, as well as, possibly, test directives steering the generation to desired scenarios. Random test generation has a history nearly as old as directed test generation, but only recently has it become the vehicle of testbench automation thanks to advances in constraint solving techniques.

Traditionally, testbenches for simulations at the RT level are written directly in design modeling languages such as Verilog and VHDL. System or behavioral level testbenches are often expressed in C/C++, which interacts with the simulator via a Programming Language Interface (PLI). Recently, specific testbench languages, including *e*, Vera, and SystemVerilog, have emerged, combining the advantages of both, as well as new capabilities such as constrained random simulation.

1.6.4 Checking Strategies

There are two general approaches to checking whether the simulation results conform to the functional specification. The first is reference-model based, the second monitor-based. In the reference-based approach, a reference model runs in parallel with the DUV, and the outputs of both are compared. This approach usually applies when a "known-good" model is available, for example, in microprocessor verification or in a regression test of incremental changes.

Monitor-based checking is common in the "white-box" verification approach, where the confirmation of the specification is observed by a set of monitored properties being true during the simulation. These properties assert certain behavior on critical design signals such as status registers, counters, and FIFOs, whereas reference-based approaches are limited to observing the behavior at the design boundary.

1.6.5 Coverage

Simulation verification faces a critical problem: the lack of good metrics able to gauge the quality of generated tests, or, in other words, the completeness of the verification. The simulation quality metric is often called the coverage model. Roughly speaking, there are two types of coverage: code coverage and functional coverage.

Code coverage, such as the coverage on lines and branches in a design description, allows a sense of completeness, since the description must be finite. The completeness, however, does not extend to the coverage of functionalities. In fact, code coverage is purely empirical and quite inadequate with regard to measuring how much of the design behavior is exercised.

A better measurement can be achieved by the functional coverage. Functional coverage asks whether or not certain design behaviors have been observed. Design behavior can be defined at multiple levels. System level behavior is often exhibited as the interaction of functional components and product functionalities. Unit level behavior is usually defined in terms of the transactions between bus masters and slaves, for example a burst or back-to-back transaction. The implementation level refers to all of the details that form the basis of correct behavior at the unit and system levels. Typical examples include FIFO tests, cross-products of FSM states, and arbitration.

Theoretically, any coverage defined over the structure or functionality of a digital design is finite. In practice, however, full functional coverage is usually unattainable due to the sheer magnitude of the state space. It is also possible that some coverage points are simply undetectable because there are unreachable design states we are not aware of *a priori*.

Therefore, simulation verification is left with no feasible stopping criterion and is deemed incomplete. A typical industry practice is to first simulate the design with a relatively small set of manually created directed tests, followed by simulating with random vectors for as long as is feasible.

Recently, with the introduction of assertion-based verification [FKL97], coverage may be expressed as a set of assertions monitored by both simulation and formal verification. The latter has the potential to prove that an intended coverage point is never reachable, thus saving simulation cycles that would have otherwise been spent in vain.

1.7   Emulation

Emulation is a technique that accelerates simulation in hardware. It maps and executes a design description in reconfigurable logic arrays, usually FPGAs. The hardware is driven by a testbench (in either software or hardware) or by the real-world environment where the design is supposed to reside. Because of the mapping, the emulated design must be synthesizable. Also, emulation models are normally only cycle-accurate and are therefore not suitable for timing verification.

Due to the intrinsic concurrent computation power of hardware, emulation is usually orders of magnitude faster than simulation However, if the model is executed off a testbench, which is traditionally non-synthesizable and therefore has to be implemented in software, the communication between the testbench and the hardware model can cause a severe degradation of the overall emulation speed.

There are several solutions. The simplest one is to load a large sequence of tests into the emulation memory, which are then executed in batch mode. In this way the communication is minimized, though the drawback is that the tests have to be pre-determined and are not responsive to the emulation.

Another solution is to have an on-board CPU handle the testbench so that communication latency is reduced. However, testbench execution by a general-purpose CPU can be a gating factor in emulation speed.

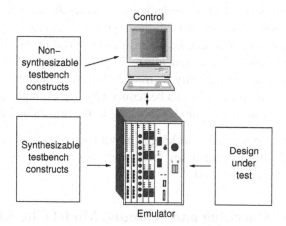

Figure 1.10. Emulation with synthesized testbench.

A recent attempt to solve this problem is to make the testbench as synthesizable as possible, so that a considerable portion of the testbench execution happens in the native hardware model (Figure 1.10). Because constraints are generally synthesizable, the constrained random generation approach to defining testbenches provides exactly the needed foundation.

1.8 Static Verification

In contrast to simulation, static verification methods, such as theorem proving [BM88, Gor88, ORSvH95, KM96] and model checking [EL85, BCMD92, Pix92, McM93, BHSV+96], and language containment [Kur87, Kur93, ASSB94, Kur94, TBK90], offer complete verification but require either extensive human expertise or enormous computational resources. This section provides a summary of formal verification methods with various levels of acceptance within the industry.

1.8.1 Equivalence Checking

Equivalence checking compares two representations of the same design. It is probably both the most popular formal verification method and the most mature, with respect to industrial practices. The common uses of equivalence checking include comparing an RTL description with synthesis results and verifying that an incremental change has not caused a functional variation. Additionally, gate-to-transistor comparison is a valuable tool for library verification. Because of its relatively low complexity, equivalence checking scales well to large designs and has gained wide acceptance.

However, the expectations for equivalence checking continue to increase with design size. A new technical challenge is sequential equivalence checking, where no one-to-one mapping of check points in the two representations being compared exists. This can be the result of optimizations such as the retiming and re-encoding of state machines. In general, sequential equivalence checking has the same complexity as model checking, though there are special cases where this complexity can be reduced by, for example, reversing the optimizations or by localizing model checking to the unmatched portions.

More recently, equivalence checking is moving to a design description level higher than RTL. In this case a key technology is the ability to reason about functionalities at the word level.

1.8.2   Model Checking and Bounded Model Checking

Model checking relies on building a finite state model and checking whether or not the desired properties hold in that model. A property, represented by a formula in temporal and modal logics [CE81, EL85, Eme90], corresponds to a set of states that can be algorithmically obtained by fixed-point analysis of the formula on the finite state model [EL86]. The set of states reachable from a designated set of initial states can also be obtained via a fixed-point analysis, commonly known as the reachability analysis. The validity of the formula is then established by the inclusion of the reachable states of the design in the set corresponding to the formula. An error trace can be generated if the property fails. Model checking is essentially an exhaustive state space search that, since the state space is finite, is guaranteed to terminate.

Model checking can be done explicitly, where the state space is enumerated, or implicitly, by representing the state space with special symbolic data structures such as a *Binary Decision Diagram* (BDD) [Ake78, Bry86]. Implicit model checking is usually more powerful [BCMD92].

Recently, the reach of model checking has been considerably enhanced by the application of greatly enhanced propositional satisfiability (SAT) solving techniques. Basic SAT-based model checking is "bounded" [BGCZ99] in the sense that a property can be shown to hold up to certain clock cycles. The effect is equivalent to an exhaustive simulation with the same bound on the number of cycles. Bounded model checking can be extended to unbounded model checking with the help of induction. Also, unbounded model checking can be done with SAT and BDD combined [MA03, GGYA03] or by employing a special approximation technique with SAT only [McM03].

1.8.3 Theorem Proving

In theorem proving, the design and the properties are expressed as formulas by using mathematical logic. A property is proved if it can be derived from the design in a logical system of axioms and a set of inference rules. The proof, resembling a combinatorial search, may appeal to intermediate definitions and lemmas in addition to the axioms and rules. Theorem proving typically requires a great deal of skilled human guidance and is mainly limited to applications at the algorithmic level. Successful case studies can be found, for example, in the verification of the Motorola MC68020 microprocessor object code [Yu92], the AMD K86's division algorithm [MLK98], and the SRT division algorithm [CGZ96].

1.8.4 Language Containment

Language containment treats the property and the design as two finite state automata. Verification is cast in terms of whether the formal language of the property automaton contains the formal language of the design automaton.

In spirit, language containment is very similar to model checking. In fact, model checking of properties in Linear Time Logic (LTL), the foundation of the now popular Property Specification Language (PSL), is modeled as a language containment problem.

Unlike theorem proving, both model checking and language containment are fully automated. However, the two share a major disadvantage, the *state space explosion* problem. In the worst case, exhaustive state exploration requires computation exponential to the number of storage elements in the design.

1.8.5 Symbolic Simulation

Symbolic simulation propagates symbols, as opposed to the constant values of conventional simulation. In symbolic forms, such as an algebraic expression, all assignments to the inputs (Boolean or integer-valued) are allowed. Therefore, all input-output combinations are captured.

This exhaustiveness classifies symbolic simulation as a formal method. For example, combinational equivalence checking can be considered a one-cycle symbolic simulation where the symbolic results are represented in some canonical form, such as a Binary Decision Diagram [Ake78, Bry86].

Like the other formal methods, symbolic simulation suffers from the state explosion problem. Various optimizations can be applied. For example, one can limit symbolic propagation to critical control logics and take the "uninterpreted function" [SGZ+98] approach to black-box memory accesses or arithmetic operations. Symbolic simulation is most useful in applications when a limited number of cycles is required, such as in memory verification [Bry91a] and pipelined implementation vs. ISA [BF89].

1.8.6   Hybrid Simulation and Formal Verification

Mixed simulation and formal verification reaps the benefits of both worlds. There many ways to combine the two.

The simplest hybrid verification is to interleave simulation and formal verification. First, a quick (constrained) random simulation is performed to grab the "low-hanging-fruit," such as bugs that can be easily identified by any verification method. Usually those are either configuration/setup errors, or bugs with shallow sequential depth. Then formal verification can be applied. More bugs may be found due to this exhaustive search. When formal verification runs out of its assigned computational resource, a massive random simulation kicks in to try covering the remaining proof targets.

A more intelligent way of doing this is to pick "interesting" states from simulation traces so that formal verification starts from those states instead of the initial state. A state is interesting if it is close, for example, measured by "Hamming-distance," to the proof targets [YSAA97], or to a set of "light-houses" [YD98, YAS00] that are deemed critical to reaching the proof targets.

Another variation is to interleave the simulation and formal search at a finer scale so that the two match in lockstep. For example, after simulating for a fixed number of cycles, a formal method, such as, SAT or BDD, is used to solve the toggling condition of a set of "tag" signals indicating whether or not the verification targets have been reached. If more than one such condition is found, the current simulation state is used as the starting point to fork multiple threads. Note that the formal method may solve over multiple clock cycles.

1.9   Constraints, Assertions, and Verification

Constraints and assertions are two faces of the same coin. Both are formal and unambiguous specifications of behaviors and aspects of a design, such as bus protocols and arbitration, as well as such implied design intentions such as freedom from bus contentions and deadlock in state machines. While assertions refer to the properties to be verified, constraints refer to the conditions required for a verification. Assertions in one verification may become constraints in another, and vice versa. This duality is exploited in the so-called "assume-and-guarantee" style of verification, illustrated in Figure 1.11.

Constraints and assertions are also the basis for two emerging verification methodologies: constrained random simulation and assertion-based verification.

It should be noted that constraints are not limited to verification. In fact, the idea of "constraint programming", the combination of the complementary strength of declarative constraints and procedural programs (e.g., Prolog), originated in the computer science and artificial intelligence field, which preceded the EDA industry. While the constraint-based verification methodology are be-

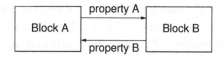

(a) Verify properties at the output of A and B together

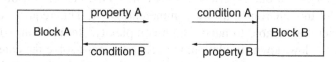

(b) Verify property A, assume property B (c) Verify property B, assume property A
   as the condition at A's input   as the condition at B's input

Figure 1.11. Verification with assume-and-guarantee.

ing widely accepted, the benefit of introducing constraints to the predominantly procedural approach to design has just begun to be realized. An early adopter, Bluespec [Blu05], for example, uses SystemVerilog assertions as integrated parts of procedural designs.

1.9.1 Constrained Random Simulation

Constrained random simulation is a key element in testbench automation, itself one of the main aspects of the recent move to high level design and verification.

A testbench is the driver of a simulation. It is used to inject inputs into a design, possibly reacting to the design's outputs, or to monitor the outputs of the design. Constrained random simulation relies on a constraint-based testbench. Constraints are formal interface specifications. These constraints can take several forms such as Boolean formulas whose variables reference inputs and state bits in the design or in auxiliary finite state machines, or in the form of temporal logic expressions. Constraints are executable through constraint solving. This fulfills an important requirement for constraints to function as a testbench.

A constraint-based testbench has many advantages over a traditional testbench. It reduces the risks due to the hand-translation of the specification into procedural descriptions, as in traditional testbenches. Randomization during constraint solving enables the simulation to explore every possible scenario allowed by the constraints. In the case of coverage-driven, or reactive, testbenches, biased randomization and additional constraints can focus the simulation further on corner cases.

Another advantage of the constraint-based testbench is drawn from the generator/monitor duality of constraints. This duality means that the very same syntax can be used to monitor the interaction between designs and to drive inputs to a design fragment.

A key to the success of constrained random simulation is the constraint solving efficiency. In our experience, sometimes hundreds of constraints are used to model the environment of a commercial DUV. This requires that the stimulus generator be able to handle high complexity. In addition, so as not to inordinately slow down simulation, the generator must solve the constraints every clock cycle very quickly, depending upon the value of the state-holding variables sampled from the DUV.

1.9.2   Assertion-based Verification

Assertion-based verification combines assertions, formal verification, and simulation, and offers the following advantages:

- It leverages power of formal verification.

- It provides a framework for hybrid simulation and formal verification.

- It facilitates IP reuse.

- With the ability to monitor internal signals and catch violations locally, it improves observability and debugging.

For a comprehensive treatment of this topic, refer to the book "Assertion-Based Design" by Foster, Krolnik, and Lacey [FKL97].

1.10   A Composite Verification Strategy

We have seen that the strength of simulation and formal verification can be combined into a hybrid verification approach, and that assertions and constrained random simulation have the potential to provide a platform for this combination. However, a few issues at the methodology level must be addressed beforehand.

Formal verification, with the exception of equivalence checking, verifies properties. Properties can be user-provided or automatically extracted. User-provided properties are not limited to assertions. They can also be scenarios in the simulation test plan that are either predetermined as suitable for formal verification (for example, properties of bus protocols), or destined for simulation without being covered at the end of the simulation.

Automatically extracted properties are usually common design practices such as no-bus-contention, no-unreachable-FSM-states, no-deadlock, and meta-stability of clock domain crossings. Checking these properties is sometimes called "super-linting." It is very likely that these properties may overlap coverage points in the simulation test plan.

Just as formal verification can check scenarios that are hard to cover in simulation, sometimes the reverse is true. This is especially the case when the design is too big for any formal verification method to handle. Properties are checked in simulation usually as *monitors* of simulation traces.

Another important issue is the definition of a common environment for simulation and formal verification. This can be handled elegantly by using constraints [Pix99]. The belief that formal verification does not need a testbench is a myth: in formal verification, the testbench consists of the constraints that capture the environment of the design being verified. Such constraints can be used directly in constrained random simulation, along with additional constraints that direct the simulation to specific scenarios. Constraints are used in simulation as generators of stimuli.

Figure 1.12 gives an example of a hybrid simulation and formal verification flow.

1.11 Summary and Book Organization

We have just given an introduction to verification challenges, as well as various approaches to addressing these challenges. A fundamental solution is the move to high level design and verification. Critical to high level verification are the two emerging verification methodologies: testbench automation and assertion-based verification. The main strength of testbench automation is its constraint-based random simulation capability. The close relationship between constraints and assertions provides a bridge between simulation and formal verification.

This book explores these ideas generally, as well as in both depth and breadth. The remaining chapters are structured as follows.

In Chapter 2, we preview the main components of constrained random simulation. A real-world example is used to demonstrate how these components are put together to form a constraint-based testbench tool. The tool's integration with assertion and formal verification is also discussed.

Chapters 3 and 4 are devoted to front-end language issues, while the remaining chapters give an in-depth treatment of the back-end technical issues of constrained random simulation.

In Chapter 3, high level design and verification languages are reviewed, together with the specific requirements of constrained random simulation language constructs. Later in this chapter, we focus our discussion on the verification language of SystemVerilog. This refines the constrained random simulation concepts introduced in Chapter 2.

Chapter 4 covers assertion languages and shows how assertions are used in simulation as result checkers, or plainly as constraints for the test generation role. These ideas are demonstrated with the Property Specification Language (PSL), along with an in-depth discussion of the language.

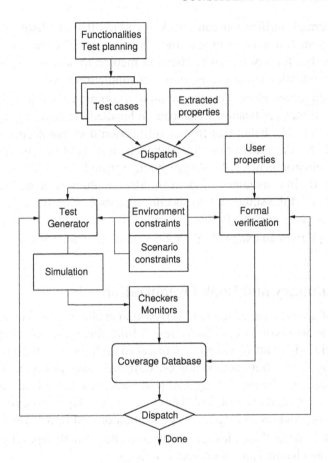

Figure 1.12.   A hybrid simulation and formal verification flow.

Chapter 5 begins a discussion of technical issues of constrained random simulation by examining common terminologies, and preliminary data structures and algorithms.

In Chapter 6, a symbolic (BDD-based) constrained random generator and related algorithms are covered in-depth. Chapters 7 and 8 continue the discussion of optimization techniques for this symbolic generator.

Chapter 9 discusses an alternative constrained random generator based on the idea of constraint synthesis. Several related approaches are also covered in-depth.

In Chapter 10, we discuss constraint diagnosis, a practical problem in constraint-based verification.

Chapter 11 provides a survey on the recent developments in word-level constraint solving. Word-level constraint solving is a natural response to the evolution of verification from bit-level (netlist) to word-level (RTL) and even higher levels of abstraction such as transactions.

The Appendix contains the list of acronyms and the proofs of the theorems and lemmas used in the book.

Chapter 2

CONSTRAINED RANDOM SIMULATION

Constrained random simulation addresses two major problems of the traditional testbench approach: being procedural as opposed to declarative and being enumerative as opposed to comprehensive. Test cases and coverage metrics are planned according to functional specifications, which are more declarative than procedural. Converting a declarative specification into test generation procedures is generally a synthesis process, one that is not particularly suitable for manual handling. Additionally when errors are made, the debugging is difficult since the specification is not executable.

The enumerative approach of traditional testbenches is also time-consuming and error-prone. Although some degree of enumeration, from coverage metrics to functionalities to be tested, is unavoidable in simulation, a simple-minded exercise of all the functionalities will not do the job because the most elusive bugs reveal themselves only under such convoluted scenarios as a series of cross states among several state machines that are set up only by a particular transaction sequence. Besides writing as many directed tests as possible, the only reasonable solution is to employ random simulation. However, the random simulation capability offered by existing HDLs is very limited.

Constrained random simulation attempts to address all of the above issues. First, as constraints are declarative, they are closer to specifications in that they mimic written language. Furthermore, because constraints are executable in the sense that stimuli are generated directly from constraints, manual construction of a testbench is eliminated. In other words, the use of constraints moves testbench writing to a higher abstraction level, one more closely resembling that of the specification. This alleviates the effort and pitfalls associated with writing a procedural-level testbench. This benefit closely resembles the advantage of logic synthesis, that is RTL to gate: higher productivity with fewer errors.

Second, constraint solving allows for flexible randomization so that case enumeration can be drastically reduced or eliminated altogether. Third, constraints can be easily converted to monitors and used at the result checking stage. This generator/monitor duality of constraints is especially useful in hierarchical verification where interface constraints for one design block are checked as properties when verifying the neighboring blocks or blocks at a higher level of hierarchy.

Failures in simulation can reveal bugs both in the specification and in the constraint-based testbench. It is therefore possible to start simulation with a fully verified testbench. Finally, in constrained random simulation, feedback from functional coverage can be used to direct constraint solving and randomization towards the not-yet-explored behavior, bringing into reality the so called "reactive testbench."

In the remainder of this chapter we discuss the main concepts of constrained random simulation, and demonstrate these concepts by describing a real constrained random simulation tool, the *Simgen* [YSP+99, YSPM01] system developed at Motorola.

2.1   Constraints for Test Generation

Constraints are formal and unambiguous specifications of design behaviors and aspects. In the context of constrained random simulation, constraints define what input combinations can be applied and when.

There are basically two types of constraints in constrained random simulation: the environment constraints and constraints used as test directives. The former defines the interface protocol, which must be strictly followed. The latter are used on top of the environment constraints to steer the simulation to the desired test scenarios, the so-called "corner cases." In other words, constrained random simulation is meaningful only if the test stimuli meet certain requirements of the environment, and can become more interesting with the help of additional test directives.

A constraint can be as simple as a Boolean formula defined over design signals. As an example, consider a typical assumption about bus interfaces: the "transaction start" input (ts) is asserted only if the design is in the "address idle" state. This can be captured in the following constraint:

$$ts \rightarrow (addr\_state == Addr\_Idle)$$

When the constraint formula gets too complicated, it may be a good idea to define intermediate subformulas upon which the constraint formula is defined. The use of macros and functions enables the sharing of common subformulas.

Auxiliary variables can be used to remember past states to constrain the sequential behavior of the inputs. The following statement from the PCI bus protocol specification is such an example.

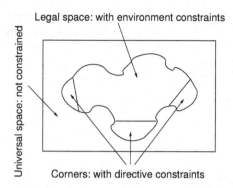

Figure 2.1.   Environment and directive constraints.

> Once a master has asserted Irdy#, it cannot change Irdy# or Frame# until the current data phase completes regardless of the state of Trdy#.

This statement corresponds to the constraint below:

    Irdy# && !data_complete -> Irdy#==prev_Irdy# && Frame#==prev_Frame#.

The auxiliary variables **prev_Idry#** and **prev_Frame#** take the values of **Irdy#** and **Frame#**, respectively, from the last clock cycle.

A class of constraints — constraints that specify bundles of sequences — can efficiently be expressed in grammar rules similar to the ones used in program language parsers. SystemVerilog allows constraints written in Backus Naur Form (BNF) to do just that. For example, the sequences

    add popf mov
    add pushf mov
    dec popf mov
    dec pushf move

can be captured with the BNF:

    sequence : top middle bottom;
    top : add | dec;
    middle: popf | pushf;
    bottom: mov;

In the most flexible form, a constraint can be facilitated by an auxiliary state machine to model complex sequential behavior.

EXAMPLE 2.1   We construct the constraint for the scenario:

> If a *request* is sent (asserted), then it will be sent continuously until after an *acknowledge* is sent.

It is understood that the *request* is an input variable and the *acknowledge* is a state variable. The scenario will be specified with a state machine and a

(Boolean) constraint expression. The state machine monitors a violation of the above scenario: a *request* is sent and then withdrawn without receiving an *acknowledge* first.

```
always @ (posedge clk) begin
    if (reset)
        state <= START;
    else begin
        case (state)
        START:
            if (req)
                state <= CHECK;
        CHECK:
            if (!req)
                state <= ERROR;
            else if (ack)
                state <= START;
        ERROR:
            state <= ERROR;
        endcase
    end
end v
```

And the constraint

```
state != ERROR
```

assures that input req is asserted in state CHECK, because this is when req was sent but /small ack has not been received, which is exactly the scenario we are trying to specify.   □

For commonly encountered constraint scenarios, the above example being one, the use of templates can save a great deal of effort. The most powerful templates are from the assertion languages. For example, the Open Verification Library (OVL) provides a set of Verilog/VHDL monitors (in state machines) that may be directly used as properties or constraints. The scenario described in the previous example can be captured with the following OVL constraint:

```
assert_window #(0, 1) handshake (clk, !reset, req, req, ack);
```

In another assertion language, the Property Specification Language (PSL), the above is written as:

```
assume always req -> req Until_ ack @ clk;
```

Unlike OVL, PSL is not a library of simulatable monitors. To use the above constraint in simulation, it must first be converted to a state machine, either implicitly if the simulator supports PSL, or explicitly otherwise.

To summarize, a constraint definition can be broken into two parts: the auxiliary modeling logic, and the Boolean expression. The modeling logic is either made up of macros, functions, auxiliary variables, and state machines, that are directly based on HDL or C/C++, or is constructed from assertion language constructs. The Boolean expression is simply a formula over signals from the design and the modeling logic. The expression evaluates to true or false. Assertion languages and constraint languages that specialize in modeling constraints are the topics of the next two chapters.

2.2 Constraining Design Behaviors

Constraints work on a DUV through synchronous composition: simply connecting the corresponding signals in the modeling logic of constraints to those of the DUV. The result is called the *product machine* of the constraints and the DUV. The direction of flow divides the signals into two groups: the signals that leave the DUV are called the state signals, whose values are set by the DUV during simulation, and the signals that flow into the DUV are called inputs. The goal of constraint solving is to provide the input assignment based on the state signal values. Figure 2.2 illustrates a constraint/DUV product machine in

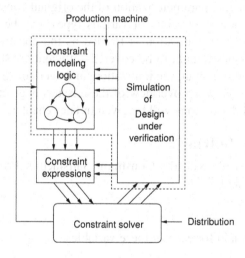

Figure 2.2.   Constraint solving.

simulation. The role of constraints can be seen as a pruner of the "raw" design behavior under unrestricted input. Test generation with constraints can be considered a process of pruning and randomization, as in the following steps:

1. *Simulation*: Apply the input from the current constraint solution, simulate, and update the current state of the product machine. (Do nothing at the beginning of the simulation when no input is available.)

2. *Constraint solving*: Propagate the current state assignment of the product machine to the constraints and "solve" the resulting constraints, that is, find

an assignment to the variables in the constraints that satisfies all the constraints simultaneously. The solution space represents the possible inputs allowed under the current state. The states that can only be reached through the prohibited inputs are therefore pruned.

3. *Randomization*: There are two possible outcomes in solving the constraints:

   a. There is no solution, which we call an *over-constraint*. In simulation, this means not a single input is allowed at the current state. The current state is therefore called a *dead-end state*. Simulation has to be aborted upon reaching a dead-end state.

   b. There is at least one solution. One input is randomly picked, according to some user specified probability distribution.

In this flow, constrained random generation is reduced to a generic constraint satisfaction problem of Boolean formulas. The propagation of state assignment can be seen as a dynamic parameterization of the original constraints.

We have shown how constraints of various types may be expressed, and how constrained random generation is formed as the constraint satisfaction problem. Several issues remain to be covered: efficient constraint solving to enhance simulation throughput, randomization, and over-constraint debugging. In the sections that follow, an overview of these issues is given and questions are raised. Detailed discussions will be given in later chapters.

2.3   Constraint Solving

Constraint solving deals with the Constraint Satisfaction Problem (CSP). A general CSP consists of

- a set of variables,

- a non-empty domain for each of these variables,

- a set of constraints restricting the values of these variables, and

- optionally, a cost function that measures the quality of assignments to these variables.

Constraint solving is to find an assignment to all of the variables that simultaneously satisfies all the constraints, or to determine the non-existence of such an assignment. With many important applications in EDA, CSPs originated and are still the subject of intense research in artificial intelligence, operational research, and database queries.

Depending on the domains being continuous or discrete, and finite or infinite, and depending on the constraints being linear of non-linear, the general CSP definition can be refined to many familiar special cases. We give a brief overview of special CSPs commonly encountered in EDA. Almost all are computationally intractable; refer to [GJ79] for more details.

- *Linear Programming (LP)* LP solves a set of linear constraints over reals and optimizes the solution according to a cost function. A special case of IP, the *Integer Linear Programming* (ILP), gives solutions in integers. Many digital logic optimization problems can be modeled as ILP, for example, Pipeline Resource Scheduling.

- *Propositional Satisfiability (SAT)* and *Automatic Test Pattern Generation (ATPG)* Both SAT and ATPG are search-based methods for solving propositional formulas in the Boolean domain. SAT works off a set of clauses that are equivalent to a gate-level design, whereas ATPG works off the design itself.

- *Binary Decision Diagrams (BDD)* BDD is a graph-based data structure that represents a Boolean function. For a fixed variable ordering in the graph, BDDs for logically equivalent functions are isomorphic. Solving a constraint is the process of constructing the corresponding BDD. The resulting BDD represents all solutions of the constraint.

- *Boolean Unification (BU) and Synthesis* A Boolean constraint can be solved via the classical BU method [W. 88, MN89]. The solution is an array of functions, each of which generates the value of a variable in the constraint. This solution method can be viewed as synthesizing a constraint into a vector function.

- *Constraint Logic Programming (CLP)* CLP combines constraint solving and logic programming. A CLP tool may solve problems ranging from linear and non-linear constraints, to propositional logics, and even temporal logic. The underlying constraint solvers can be logic-based (e.g., SAT), mixed LP and ILP, and interval arithmetic, over the real, finite, and Boolean domains.

When used in constrained random simulation, ILP, CP, SAT, and ATPG are classified as *on-line* approaches in the sense that the solution happens during simulation. BDD and BU are off-line because the solution happens when the BDDs and solution functions are constructed, usually before simulation. The trade-off is that the on-line approaches take exponential time in the worst case, while the off-line approaches use exponential space in the worst case. As we will see later, these exponential complexities can often be avoided with heuristic data structures and algorithms.

With the move to a higher model level, constraints are routinely expressed at the word-level instead of at the Boolean level. For example, transaction requests and grant signals have as many scalar variables as there are bus masters.

However, they are referenced as single word-level variables. While ILP and CP are natural word-level solvers, the others are originally designed to work with Booleans. The most efficient approach, however, is the combination of both.

2.4   Efficiency of Constraint Solving

One important characteristic of simulation is that it scales up to large designs. This is its main advantage over static verification. Complex designs, however, come with a large number of constraints. Even though constraints are concise, the complexity of constraint solving can easily be multiplied by the hundreds or thousands of constraints and variables when verifying a unit- or even a block-level design.

Fast constraint solving is critical to high simulation throughput and, therefore, good coverage. Enhancement of constraint solving efficiency can come from many sources. We will discuss these enhancements in later chapters; examples include:

- Choosing the right solver and improving its capacity. For example, bit-level versus word-level.

- Simplifying the problem with the partitioning, decomposition, and extraction of special constraints.

- Changing the problem with prioritized constraints and soft constraints.

2.5   Randomization

Besides giving the correct inputs, giving "good" inputs that are more likely to exercise interesting scenarios is a top priority for a constrained random generator. Choosing good inputs from correct ones is not as easy as it sounds. There are a few questions that must be asked when determining good inputs.

First of all, what is a good choice? In most cases, a good choice means that every possible input gets an equal chance to be selected, making the simulation effort spread over the entire input space. This fairness assumes that every input has the same weight in filling up the coverage chart, which is not always the case. For example, in generating data inputs to arithmetic operations, it is better to give much higher weights to the minimum and maximum values than to the ones in between. This ensures that boundary conditions are sufficiently exercised.

Second, given the desired distribution or weighting, how do you implement it? This should be achieved under the reasonable assumption that constraint solving and randomization are done simultaneously.

Related to these questions, how does one to make randomization adaptive to simulation?

Finally, how does one maintain *random stability* so that simulation traces may be repeated in identical settings? This is very critical to debugging.

2.6 Constraint Diagnosis

The problem of unsatisfiable constraints, or an over-constraint, has serious consequences in both simulation and formal verification. Over-constraints stall the progress in simulation and lead to vacuous proofs in formal verification. Constraint diagnosis identifies the source of and helps to fix over-constraints.

2.7 A Constrained Random Simulation Tool

In this section, we discuss a real-world tool, *Simgen* [YSP+99]. *Simgen* stands for simulation generation. It provides a constrained random simulation framework. *Simgen* is written in C++ and interfaces with a Verilog simulator through PLI. It recognizes a small set of syntax for defining constraints, randomization, and simulation control.

Simgen also works with a formal verification tool and an assertion language in an integrated dynamic and static verification environment. Line and branch coverage are given for the assertions, and state and transition coverage are given for any state machines the user specifies. Figure 2.3 gives a flow chart of *Simgen* and the integrated verification environment.

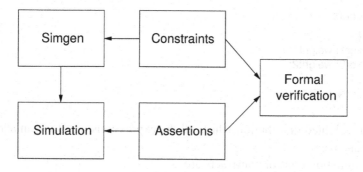

Figure 2.3.   Simgen, simulation, and formal verification.

2.7.1 The Language

The *Simgen* constraint and randomization language follows the conventions of Verilog system calls. However, constraints and randomization are handled in preprocessing by *Simgen*'s Verilog compiler. Only the simulation controls are really system calls.

A constraint is declared with the syntax

```
$constraint (name, expr);
```

where name is a label for the constraint, and expr a Verilog Boolean expression that is asserted by the constraint. The name provides a handle to turn the constraint on and off statically (before simulation).

By default, a constraint is applied at every edge of a fast virtual clock. The user can also specify at what clock the constraint should be applied using

```
$constraint @ (posedge|negedge clock) (name, expr);
```

Clocks can be generated with clock constraints

```
$clock_constraint (name, expr);
```

which usually have the need to call the facility function

```
$prev(var)
```

to fetch the value of variable var from the last virtual clock cycle.

Randomization can be done in two ways: assigning weight to input variables, or to Boolean expressions.

```
$setprob1 (var, weight); $setprob0 (var, weight);
```

The above statements define how likely the input variable var will take the value 1 or 0, respectively. The likelihood is assigned to weight, which is a real number or a Verilog expression that evaluates to a real number between 0 and 1. Here is an example:

```
$setprob1 (x, state == IDLE ? 0.9 : 0.5);
```

The syntax

```
$dist {
     expr1: weight1;
     expr2: weight2;
     ...
     exprN: weightN;
}
```

defines a weighted distribution where all the expressions are Boolean and mutually exclusive.

The simulation control constructs are

```
$simgen_stable;
$simgen_start;
$simgen_finish;
```

where the first system call is used to detect the stability of the simulator (usually the end of initialization), the second starts the *Simgen*-driven simulation, and the third performs post-processing.

Simgen uses one random generator. Its seed can be set with the following syntax.

```
$simgen_seed (seed);
```

Simulations can be reproduced if they start with the same seed and if the other settings remain the same.

2.7.2 BDD-based Constraint Solving

Constraints are first compiled into BDDs and the conjoined. A special BDD traversal algorithm is used at simulation time to generate vectors. The generator has the following properties:

1. If the constraints are satisfiable, then the generator will produce an input with no backtracking.

2. Without user-specified weighting, the generated inputs follow the uniform distribution.

Constraint solving in *Simgen* is preceded by several passes of optimizations, including extraction and simplification, partitioning, and decomposition [YKAP02, YPAA03].

2.7.3 Prioritized Constraints

To further enhance its constraint solving capacity, *Simgen* introduced a new concept called *prioritized constraints*. Prioritized constraints are basically normal constraints assigned with priority levels. The higher priority constraints are solved first, and the solution is propagated to the lower priority constraints before they are solved.

One characteristic of this approach is that input variables are not solved simultaneously. Rather, they are solved in an order implied by the hierarchy of the constraints. This alters the semantics of the original constraints by making them tighter. However, situations of sound prioritization abound. To take a simple example, clock constraints should always be solved before other constraints. Also, constraints regarding the transaction type can safely be solved before the constraints concerning transaction attributes.

Chapter 3

HIGH LEVEL VERIFICATION LANGUAGES

High Level Verification (HVL) language is a relatively new concept; not long ago most verification activities happened in HDLs. Previously, high level verification was done using C/C++ behavioral models, which had no direct connection to hardware. As the verification problem began getting out-of-hand at an alarming rate, people realized that a methodology change was imminent. But before this could happen, a language that specifically addressed verification needs was required.

The demand is so strong that multiple verification oriented languages have been developed in the span of just several years. For testbench automation, there are Verisity's *e*, Cadence's TestBuilder, Synopsys's Vera, and now SystemVerilog, just to name the major players. For assertion-based verification, we have OVL and PSL; *e*, Vera, and SystemVerilog each has its own assertion language.

Parallel to the development and deployment of HVL, SystemC and SystemVerilog are being drafted, so it makes economic sense for the existing HVLs to become part of the system languages. As a result, SystemVerilog's random constraint language (SVRC) derives directly from Vera, donated by Synopsys to the language standardization organization, Accellera. SystemC Verification library (SCV) is a SystemC-flavored Testbuilder implementation.

Another verification component is the assertion language. SystemVerilog takes a similar move by adopting OpenVera Assertion, also a contribution from Synopsys, which becomes the SystemVerilog Assertion (SVA). SVA was modified to align it with other languages. Accellera also endorses PSL, yet another assertion language. PSL can be written in many flavors, including Verilog, VHDL, and SystemVerilog. SystemC-flavored PSL is being developed.

Assertions are not only properties to be proved, but they also have a deep involvement in testbench: they serve as the checkers or monitors in functional coverage, and they also serve as constraints that define the input behavior in verification. Actually, the assertions in OVL, PSL, and SystemVerilog can all be used as "assumptions." In this sense, these assertion languages provide a way of modeling the test environment for both simulation and formal verification.

This chapter is intended to be a discussion of the key concepts in HVL rather than a language reference or a tutorial. We start with a summary of features in the two system languages, SystemVerilog and SystemC. We then give a comparison, along with the language *e* from Verisity, of their verification and testbench languages. After that, we focus on the SystemVerilog testbench language and its main ingredient, the constrained random simulation. We will cover assertions and constraints in the next chapter.

3.1   Testbench and System Languages

Traditionally, testbenches are written in HDL, that is, Verilog or VHDL. An important aspect of a testbench is its interface with hardware. Therefore, the HDL's hardware signal-level modeling and concurrent execution is indispensable. For higher level testbench behavior, such as control and test case generation, HDLs do provide limited support. For example, Verilog provides for loop, fork/join, integer types, and the $random() system function. Furthermore, HDLs do support basic forms of Object-Oriented Programming (OOP), such as (parameterized) module and (parameterized) instantiation. The difference is that the instantiation of modules is static, whereas in C/C++ one can have dynamic objects. A complete OOP solution also provides a rich set of features that are desirable to programmers of all kinds, including hardware design/verification engineers. These features include data encapsulation, inheritance, polymorphism, overloading, and virtual functions.

Leveraging programming language standards and know-how for testbenches has been common practice since the early days of HDL. For example, Verilog offers the Programming Language Interface (PLI) so that C/C++ routines can be invoked as callbacks from the simulator. The VCS simulator provides an enhanced version of PLI, called DirectC, which compiles C/C++ routines together with the design.

Naturally, high-level data structures common to software programming languages, such as array, list, and queue, have also been adopted to HVL. Modeling hardware behavior above the RT level requires a generic concurrency model that includes synchronization and communication.

Therefore, our items on a wish-list for the next generation of design and verification languages would include:

- Adequate RTL modeling capacity

- Object-oriented features

- High-level data structures

- Concurrency

- Testbench automation

- Assertion capacity

Both SystemVerilog and SystemC seem to meet these requirements. SystemVerilog is a superset of Verilog2001. In addition to its industry standard HDL capacity, SystemVerilog adopted many features from SuperLog. It also took the donation of testbench/assertion languages and the DirectC programming interface from Synopsys. As for high level programmability, SystemVerilog allows flexible data types such as strings and events, as well as classes that can be dynamically created and deleted. Figure 3.1, taken from Goering's EE Times March 2002 article [Goe02], lists the main constructs and data types of SystemVerilog and illustrates its relation to Verilog2001 and C.

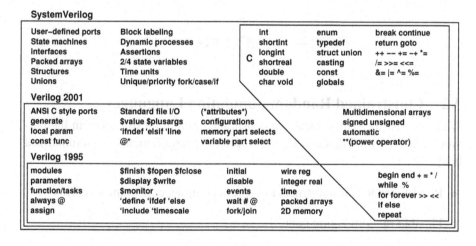

Figure 3.1. SystemVerilog features.

In contrast to SystemVerilog's "from hardware to software" approach, SystemC takes the other direction. SystemC is a C++ library equipped with hardware modeling capabilities. For example, the core language defines events, processes, modules, and ports. Application specific libraries can be added on top of the core but are considered separate from the SystemC language. A recent addition to the SystemC verification library SCV, a SystemC-flavored Testbuilder implementation, addresses a shortcoming in testbench automation. There are plans for it to adopt the PSL assertion language. What makes SystemC unique is that it provides a reference event-driven simulator. Figure 3.2 gives the main components of SystemC.

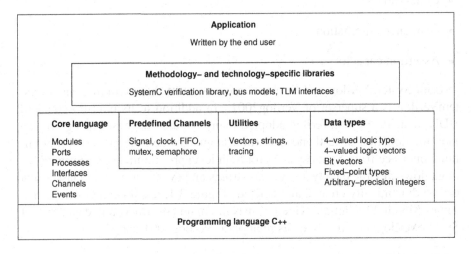

Figure 3.2.   SystemC language architecture.

3.2   Constrained Random Simulation Languages

We discussed the requirements of constrained random testbenches in the previous chapter by using a concrete example. The requirements are summarized below.

- Robust syntactical constructs to express complex constraint scenarios

- Flexible randomization controls

- Dynamic constraints and randomization

- High performance constraint solver

We will now compare three constraint-based testbench languages, SVRC, SCV, and *e*, for their conformance to the first three requirements. For a quick overview of the compared features, refer to Table 3.1.

3.2.1   Constraints

Since all three languages work with HDL (SVC through SystemC's *Models of Computation*), Boolean constraints with all the logical and relational operations are already very expressive. In addition, all three languages introduce set and membership functions that are commonly used in test generation. SVRC further added control constructs such as *if/else*, *case*, and inline constraints, and is the only one that provides grammar-production-based sequence generation. SCV and *e* allow "soft" constraints – constraints that are considered only if they do not cause the constraint solving to fail.

3.2.2   Commutativity of Constraints

The commutativity of constraints affects both how constraints are solved and randomization. A constraint is commutative if all random variables are solved simultaneously. In all of the languages, constraints are commutative by default. However, in SVRC, a variable ordering can be imposed. This provides an extra control on randomization, as well as simplifies constraint solving. The caveat is that variable ordering **alters** a constraint: it makes the constraint tighter and can cause the constraint solving to fail.

3.2.3   Randomization

Several randomization schemes are provided by these languages. By default, the SVRC and SCV constraint solvers realize a uniform distribution. That is, all legal input combinations are equally probable. All three languages use weighted (range) distribution to bias the randomization. For random sequencing, they allow permutation and non-duplicate iteration of a random variable's range of values.

Randomization in SVRC and SCV guarantees thread and object stability. That is, random value sequences are independent of thread ordering, and each constraint class object has its own random value sequence that is shielded from the randomization in other objects.

3.2.4   Dynamic Constraints and Randomization

Dynamic constraints come for free in all three languages, since constraints can be state-dependent. The state is evaluated in simulation before constraint solving, and thus "conditions" the constraint. SVRC uses additional syntactical rules to recognize "guarded" constraints. A guarded constraint is applied only if its guard evaluates to true. Also in SVRC, constraints can be explicitly disabled and enabled, and constraints can be added inline and used just once when a specific object (of a constraint class) is solved. In *e*, soft constraints can be disabled.

Like dynamic constraints, dynamic randomization can also be achieved with weights that are state-dependent. This is the case for all three languages. In SVRC and SCV, additionally, randomization can be controlled by: 1) disabling and enabling randomization of certain variables or fields of variables, 2) changing the randomization seed, and 3) using the pre- and post-processing routines.

Table 3.1.   Constraint and randomization features of SVRC, SCV, and *e*.

	SVRC	SCV	*e*
Constraints			
Boolean formulas	✓	✓	✓
set membership	✓	✓	✓
soft constraints		✓	✓
inline constraints	✓		
random sequences	✓		
if/else/case	✓		
Commutativity			
commutative by default	✓	✓	✓
variable ordering	✓		
guarded constraints	✓		
Randomization			
unconstrained randomization	✓	✓	✓
uniform distribution	✓	✓	
permutation	✓	✓	✓
non-duplicate iteration	✓	✓	✓
weighted distribution	✓	✓	✓ (soft)
weighted range	✓	✓	✓ (soft)
weighted case	✓		
pre- and post-processing	✓	✓	
Random stability			
thread	✓	✓	
object	✓	✓	
explicit seeding	✓	✓	
Dynamic constraints			
disable/enable	✓		✓ (soft)
state-dependent constraints	✓	✓	✓
guarded constraints	✓		
inline constraints	✓		
Dynamic randomization			
disable/enable	✓	✓	
state-dependent weight	✓	✓	✓
explicit seeding	✓	✓	
pre- and post-processing	✓	✓	

3.3   SystemVerilog Random Constraints

In this section we cover the language constructs of SVRC. We begin with an overview of how constrained randomization takes place in a SystemVerilog environment. Following that, we describe how constraints and randomization are specified with the SVRC language.

3.3.1 Overview

In SVRC, constraints and randomization are typically grouped into classes where each class contains some variables to be randomized, as well as constraints used in the randomization. Objects of these classes can later be extended, inherited, constrained, overridden, enabled, disabled, merged with, or separated from other objects.

A class may also contain other variables whose values are predetermined by the state of the design. Those are generally referred to as the state variables. Randomization is invoked as a default member function of the objects. All random variables and constraints in an object are solved together by default, unless one specifies a solve order for the variables.

EXAMPLE 3.1 This example shows a constraint class and the randomization of its object.

```
class Bus;
rand bit[15:0] addr;
rand bit[31:0] data;
bit[2:0] trans_type;

function new(bit[2:0] type);
    trans_type = type;
endfunction

constraint word_align {
    (trans_type != IO) -> addr[1:0] == 2'b0;
}
endclass

Bus bus1 = new(IO);
if (bus1.randomize() != 1)
    $display("Randomization failed.");

Bus bus2 = new(READ);
if (bus2.randomize() != 1)
    $display("Randomization failed.");
```

The following are the points of interest in this example:

- The class Bus has two random variables, addr and data, and one state variable, trans_type. Only the two lowest bits of addr are constrained.

- Two objects of Bus are created with different arguments: bus1's transaction type is IO and bus2's is READ.

- Randomization is invoked as a built-in method, randomize(), of an object. The state variable trans_type is first evaluated and then plugged into the constraint. Note that "->" is the logical implication operator.

- If a randomize call succeeds, it returns 1. Randomization can fail if the constraints are not satisfiable. This is not the case in this example.

- For bus1, since the trans_type is initialized to IO, the constraint is discharged; therefore, both addr and data are unconstrained. In this case, they follow the uniform distribution over their full declared range.

- For bus2, the constraint forces addr[1:0] to be 2'b00 − addr[15:2], and data again follows the uniform distribution.

                                                                                        □

Constraint and randomization can also take place outside of any class context, in an arbitrary scope. This can be done with the scope randomize function, std::randomize(), and an inline constraint invoked using the keyword *with*, as in the example:

```
task stimulus( int length );
    int a, b, c;

    std::randomize( a, b, c ) with { a < b ; a + b < length };
    ...

endtask
```

Another point from this example is that all the random variables a, b and c, and both constraints, a < b and a + b < length, are solved simultaneously.

The example below shows an instance of inheritance, using the keyword "extends".

```
class MyBus extends Bus;
rand bit addr_type;

constraint addr_range {
    (addr_type == LOW ) -> addr inside { [16'h0 : 16'hEFFF] };
    (addr_type == HIGH) -> addr inside { [16'hF000 : 16'hFFFF] };
}
endclass
```

Like a C++ inheritance, class MyBus inherits all the variables, methods, and constraints of Bus as defined previously, and adds both a new random variable and a constraint. Note that "inside" is the set membership function and that [16'h0 : 16'hEFFF], for example, defines a range.

The rest of this section provides more details on the language aspects of constraints and randomization mechanisms.

3.3.2 Set Membership

Set membership is a special constraint where a given expression's allowed valuation is enumerated as a range of values. Depending on the membership type (*inside* or *not inside*), the expression either must (for *inside*) or must not (for *not inside*) take a value from the specified range. In the absence of other constraints, all values inside (or outside) the range are equally probable. The example below shows the use of inside and not inside.

```
rand integer x, y, z;
rand bit[2:0] w;

constraint c1 { x inside {3, 5, [9:15], [y:2*y], z}; }
constraint c2 { !(w inside {1, 2, 4}); }
```

Note that [9:15] stands for the inclusive range from 9 to 15. Note also that, values and ranges in the set can be specified with the expressions z and [y:2*y]. This provides a mechanism for dynamically altering the constraint.

The valid range for w in the above example is {0, 3, 5, 6, 7}, which is the complement of the given range with respect to the full range of the 3-bit signal w.

Distribution over a weighted range can be used to approximate more complex distributions. Figure 3.3 shows the step-like approximation of a Binomial distribution. In the straightforward translation to an SVRC weighted range distribution, steps correspond to ranges, and probabilities correspond to weights.

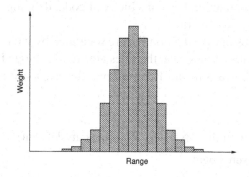

Figure 3.3. An approximation to a binomial distribution.

3.3.3 Inline Constraints

Constraints can be added inline when the randomize() function is called, using the syntax "randomize() with { <constraints>}." Depending upon how randomize() is invoked, there are two situations:

1. If it is called by an object, then the inline constraints are added to the constraints of that object.

2. Otherwise, (randomize() is called in the current scope but not from any object) the arguments of the call, for example, a in randomize(a), are treated as random variables, and the inline constraints are the only ones applicable.

The following example illustrates the two invocations of inline constraints:

```
// restrict data values to powers-of-two (one-hot)
bus1.randomize() with  { data & (data-1) == 0; } ;
```

```
int a, b, c;
std::randomize( a, b, c ) with  a < b ; a + b < length ;
```

3.3.4   Random Sequence Generation

SVRC's sequence generation allows one to specify the stimulus sequences using the BNF notation, which is a common syntax for the formal specification of programming languages. The BNF of a language is a set of production rules, each of which is of the form:

```
prod_name : prod_list;
```

where prod_name is the name of the production rule, and prod_list is a list of production names, each referring to another production rule defined elsewhere. This deductive definition stops with rules that are called terminals. A terminal rule has an empty production list and a block of code, defining the action taken when the parser reaches this rule.

The multiple rules in a production list are separated by white space or by the OR operator (|). With white space, the rules are streamed together in sequence. With |, a random choice is made. For example, this sequence of words

```
dog bites man
```

can be recognized with the BNF (terminal rules are left out)

```
sentence : noun verb noun;
noun : man | dog;
verb : bites;
```

in a bottom-up fashion so that each of the lower level rules, "noun" and "verb" are recognized, followed by the top rule, "sentence."

Sequence generation is the reverse procedure: instead of recognizing a sequence using BNF, the BNF rules are used to construct a sequence top-down, starting with the top-most rule. In the above example, a total of four sentences can be constructed starting from the "sentence" rule:

```
dog bites man
dog bites dog
man bites man
man bites dog.
```

Arguably, only the first three sentences make sense (while the last one makes news). Either the rules need to be refined or some additional constraint has to be added to make the correction.

In essence, BNF-based generation can be viewed as a procedure of unrolling a set of templates. This bears some similarity to sequence generation from regular expressions and formal assertions, which can also be thought of as a set of templates. We will discuss these issues in Chapter 4.

An SVRC sequence generator is declared with the predefined class "randseq," as in the following:

```
randseq (production_name)
{
      prod_rule1;
      prod_rule2;
      ...
      prod_ruleN;
}
```

The following is a complete random sequence generator that prints the sentences above.

```
randseq (sentence_generator)
{
      sentence : noun verb noun;
      noun : man | dog;
      verb : bites;
      man : { $display("man "); };
      dog : { $display("dog "); };
      bites : { $display("bites "); };
}
```

3.3.5   Variable ordering

By default, constraints are commutative, meaning that they define bidirectional relations among random variables. For example, to satisfy the constraint (a -> b), one can assign a to 1 and b to 1, or one can assign b to 0 and a to 0. This bidirectionality is indicated by the freedom of choosing the order of solving the variables, or, in other words, the variables are solved simultaneously.

Sometimes one may find this default behavior undesirable, as can be seen from the following example:

```
constraint c1 (a -> b == c);
```

Suppose b and c are two 32-bit signals. If all three variables are solved simultaneously, then a solution where a is 1 has a chance of 1 out of 2^{32} to be generated. If a happens to be a critical branching condition, then the chance of enabling the branch is practically zero.

The solution to this problem is to impose partial orders among the random variables. In the above example, if we solve a first, then we will have a 50% chance of seeing the branching. In SVRC, the syntax for defining an order is explicitly defined with the keywords "solve ... before ...", as illustrated in the following example:

```
constraint c1 (a -> b == c);
constraint c2 { solve a before b; }
constraint c3 { solve a before c; }
```

The semantics of the variable ordering solve a before b is as follows: in the same solution space of the constraints without the ordering, choose the assignments to some variables before choosing the assignments to other variables. Hence, variable ordering affects randomization but not the satisfiability of constraints.[1]

However, note that the specified variable order may be overridden as long as this does not change the outcome of constraint solving. Consider the following example:

```
s -> x == 0;
x < y;
solve y before x;
```

If s, a state variable, evaluates to 1, then x has to take the value 0 regardless of any variable ordering. Therefore, if we can recognize those situations and assign the only possible values first, the rest of the constraints can be solved more easily. Later in Chapter 7 we will show exactly how to recognize and take advantage of those situations.

Variable orders are transitive. That is, "solve a before b" and "solve b before c" implies "solve a before c." This can lead to a circular order, which is prohibited by SVRC. The following restrictions apply to variable ordering:

1. Only random variables of integral values can be ordered

2. Only random variables declared with "rand" can be ordered

3. No circular ordering is allowed

In particular, the "cycling Random Variables" declared with the keyword "randc" are disallowed in variable ordering because they already imply an automatic variable ordering. The following section contains a more detailed discussion of "randc".

---

[1] Vera allows a "hard" version of variable ordering, "solve a before b hard", which does change the satisfiability of constraints by selecting a subset of constraints to solve for each ordered variable. This can cause the constraints to fail, but is also a powerful tool for enhancing the constraint solving capability. Refer to Chapter 2 for hierarchical constraints.

3.3.6 Cycling Random Variables and Permutations

The keyword "rand" declares a random variable to be solved. The keyword "randc" declares two things: a random variable and that the variable is randomized via a special mechanism called "random-cyclic." The variable cycles through all of the values in a random permutation of the variable's range. An initial permutation is selected to start the cycling. At the end of each cycling, another permutation is selected and so on.

One subtlety is that the randc variables are solved for prior to the rand variables, using only constraints that involve one randc variable. As previously discussed, this gives a special instance of variable ordering. The values for the randc variables are used as constants, as is the case for non-random variables, while solving for the remaining rand variables. Consider the following example:

```
randc reg[3:0] x, y;
rand integer z;
constraint C
{
    z == x+y;
    x < 10;
    y > 5;
}
```

The second and third constraints are solved first, separately. Therefore, x will cycle over permutations of 0 - 9, and y will cycle over permutations of 6 - 15. By using the values of x and y as constants, the first constraint is solved last.

This concept of ordering variables turns out to be crucial in constraint solving, especially in enhancing the controllability of distributions. On the other hand, since an extra ordering restriction is imposed, the solver may fail to solve constraints which would otherwise be satisfiable. Note that in contrast, orderings specified with "solve ... before ..." preserves the satisfiability of the constraints.

3.3.7 Guarded Constraints

Guarded constraints are special *conditional constraints*. Conditional constraints are constraints in the following three forms:

- A -> B,

- if (A) B,

- if (A) B else C.

The consequence B (or C) applies only if the condition A is true (or false). The advantage of conditional constraints is that if the condition can be easily checked, there is a chance that the whole constraint is discharged (provided the condition is false). However, evaluating the condition may not always be possible without solving the whole constraint if, for example, the condition involves a random variable.

A guarded constraint is a conditional constraint whose condition can always be evaluated independently of the constraint itself. For this to hold, the condition expression can only consist of all of the following:

- constants

- state variables

- comparisons involving only object handles and NULL

The last form alludes to another purpose of guarded constraints (in addition to constraint solving efficiency): the protection against runtime errors of dereferencing NULL object handles.

3.3.8   Distribution

A distribution is a weighted membership. For example:

    x dist { 100 := 1, 200 := 2, 300 := 5 }

specifies that x is equal to 100 or 300 with a weighted ratio of 1-5. And another example is:

    x dist { [100:102] :/ 1, 200 := 2, 300 := 5 }

where x is equal to one of 100, 101, 102, 200, or 300 with a weighted ratio of 1/3-1/3-1/3-2-5. Note that if the weight is not given, then every value in the range has equal probability. In this case, the distribution is exactly like a membership. That is

    x dist { [100:102], 200, 300 }

is equivalent to

    x inside { [100:102], 200, 300 }.

3.3.9   Pre-processing and Post-processing

The call to the function randomize() automatically invokes two built-in functions: pre_randomize() and post_randomize(), executed immediately before and after the randomization.

These pre- and post-processing functions can be overridden to perform customized functions. Typically, pre_randomize() is used to assign values to state variables in the constraints to be solved, providing a mechanism to dynamically "configure" the constraints. The following is an example of pre_randomize():

```
class Package;      rand int length;
    bit mode;

    constraint package_size {
        mode == SMALL -> length inside { [0:99] };
        mode == LARGE -> length inside { [100:200] };
    }

    function void pre_randomize(bit m);
        super.pre_randomize();
        mode = m;
    endfunction
endclass

Package pkg = new;
pkg.randomize(SMALL);
```

Note that the call to the super class's pre-processing method, the function "super.pre_randomize()", is a standard usage. In addition to pre-conditioning constraints, pre_randomize() can also alter the randomization directly with the help of internal or external random generators. The following example defines a pre_randomize function that assigns a variable length appearing in some constraints. Two external random generators are used to generate binomial and exponential distributions. The default generator is the built-in $urandom_range().

```
function void pre_randomize(int algorithm);
    super.pre_randomize();
    if (algorithm == BINOMIAL)
        length = binomial(MAX_LENGTH);
    else if (algorithm == EXPONENTIAL)
        length = exponential(MAX_LENGTH);
    else
        length = $urandom_range(MAX_LENGTH);
endfunction
```

The built-in post-processing method, post_randomize(), is mainly for cleanup, result checking, or computation that depends on the result of the randomization. The usage is similar to that of pre_randomize().

3.3.10 Random Stability

The SVRC system randomization calls, $urandom() and $urandom_range(), are local to threads. If multiple threads are forked, the random generations in the threads are mutually independent and are insensitive to the thread ordering. This is called thread random stability.

Likewise, the object randomization call, randomize(), is local to each object, and random generations in different objects are also mutually independent. This is called object random stability.

Random stability allows the user to observe identical simulation traces in repeated runs with identical settings. This is very helpful in debugging the simulation.

3.3.11   Dynamic Constraints and Randomization

SVRC offers several ways to dynamically modify constraints on randomization:

- Conditional constraints using implication and if-else.

- Constraint objects can be turned on and off using the constraint_mode() function. Initially, all constraint objects are on.

- Random variables can be made active or inactive using the rand_mode() function. Initially, all rand and randc variables are active.

- The weights in a distribution constraint can be state-dependent.

3.4   Summary

We have compared the features of three popular testbench languages, the SVRC of SystemVerilog, the SCV of SystemC, and *e* from Verisity. The comparison showed that there is a common set of testbench constructs, centered on constraints and randomization. We have also studied the language features of SVRC as an example. Many of these features, including variable solve order, weighted range distribution, guarded constraints, and cycling random variables, will be revisited later chapters regarding constraints solving and optimization.

Chapter 4

ASSERTION LANGUAGES AND CONSTRAINTS

We have seen how testbench automation languages facilitate testbench writing and test generation. Likewise, assertion languages play the same role in assertion writing and verification. The role of a specific language is to encode commonly encountered scenarios into a small set of constructs subjected to strict syntax and semantics. For example, a complex sequence generator can either be coded in HDL or, succinctly, as a set of grammar rules in Vera. Design behaviors are often expressible in terms of sequences of events that are logically and causally related. Assertion languages are designed to capture such behaviors. What differentiates assertions from a reference model in HDL is that the latter is procedural, whereas the former is declarative. It follows that assertions share the same benefits of constraints in testbenches, such as closeness to natural design specification, conciseness, and formalism.

In many occasions, assertions and constraints are used side-by-side or even exchangeably. For example, in constraint-based testbench languages such as *e*, VERA, and SVC, assertions are used as trigger events and functional coverage points; in assertion-based verification, assertions can act as properties to be proved or as constraints that specify the verification environment.

In this chapter, we will first give a brief introduction to mainstream assertion languages, and then focus on the assertion language PSL. In describing PSL, we will begin by introducing its key ingredients, such as temporal logic, regular expression, and the strong/weak satisfaction of properties. These form the foundation of the main language constructs in PSL, which are covered next. Finally, we will discuss the assertion-to-checker conversion that is key to the monitor/generator duality of assertions, that is, assertions as monitors in simulation and static verification, and as constraints in testbenches for test generation.

4.1 Assertion Languages

With the advent of formal verification within major semiconductor design houses, followed by adoption by the EDA vendors, a plethora of assertion languages have emerged in the last decade. The major ones include IBM's Sugar, Motorola's CBV, Intel's ForSpec, and Synopsys' OVA (jointly developed with Intel). OVA is also used to define sequential coverage events in Synopsys's testbench automation tool Vera. Verisity's *e* uses assertions for the same purpose.

Table 4.1. Assertion Languages.

Abbreviation	Full name	Affiliation
CBV	Cycle-based Verilog	Motorola/Freescale
e	the *e* temporal language	Verisity/Cadence
ForSpec	ForSpec	Intel
OVA	Open Vera Assertion	Synopsys, Accellera
OVL	Open Verification Library	Verplex, Accellera
PSL	Property Specification Language	Accellera
Sugar	Sugar	IBM
SVA	SystemVerilog Assertion	Accellera

4.1.1 PSL

Since verification is a key component of the higher-level design abstraction advocated by the industry, the Accellera EDA standards organization has been seeking to end the proliferation of assertion languages. So far, two languages have been selected: PSL, which is a Sugar derivation, and SVA, which combines features from OVA and CBV [HW05].

Rooted in formal verification, PSL and SVA share core language components, especially the temporal logic aspect of the languages. However, the two still differ in some key aspects. The following comparisons are taken from [HW05].

- **User perspective**

    PSL is a stand-alone assertion language with modeling flavors to allow it to work harmoniously with Verilog, VHDL, SystemVerilog, and General Description Language (GDL). SVA is an embedded assertion capability of SystemVerilog. It can be written natively as a part of SystemVerilog designs and testbenches.

- **Language-theoretic Perspective**

  PSL is divided into the Foundation Language (FL) and the Optional Branching Extension (OBE). FL is a mixture of Linear Time Logic (LTL) and an extended Regular Expression (RE). FL is a linear temporal logic that reasons about computation paths. OBE is a branching temporal logic that reasons about computation trees, that is, either multiple paths at once or, optionally, some of the paths, one at a time.

  As a whole, SVA is a linear temporal logic that is comparable to the FL of PSL. It is based on *sequences* which are extensions of REs. Above the RE level, however, FL supports the full LTL constructs, whereas SVA directly provides only the "always" construct, while indirectly supporting the *weak* constructs of LTL through the recursive definition of assertions. As a result, FL can express both the *safety* and *liveness* assertions, while SVA is oriented more towards safety.

- **Data Manipulation in Assertions**

  Within the definition of an SVA assertion, local variables can be created and assigned to remember past states of the execution. This is a powerful facility for the robustness of assertions. PSL does not support local variables directly. A limited version of local variables can be found in the use of the "forall" construct.

The fact that two similar assertion languages have been selected by the same standard organization can be confusing, but the fact that both are endorsed standards has also encouraged the spread of formal verification. In addition, Accellera has coordinated a PSL/SVA alignment effort [DEF^{+}04] in which a common kernel of both languages is identified and aligned. Syntax and semantics changes are made to both PSL and SVA so that, for the same semantics, the same or similar syntax is used as much as possible. As a result, a large subset of SVA can be translated to PSL, and vice versa [HFEM03].

4.1.2 OVL

Given its important role in the deployment of assertion-based design, OVL deserves a quick introduction. As it reads, OVL, the Open Verification Library, is not really a language but a library of commonly encountered low-level assertions, related to buses being one-hot, signals being stable, and requests being acknowledged, etc, built from two hosting languages. The hosting languages, not surprisingly, are the popular Verilog and VHDL. This makes OVL assertions both compatible with most HDL simulators and ready to be checked by static methods. Prior to the development of PSL and SVA, OVL played an indispensable role in introducing the concept of assertion to engineers in the languages they were already familiar with. In this sense, OVL has been a successful idea.

The standardization of PSL and SVA has been a mixed blessing to OVL. With more vendors supporting these standards natively in HDLs, the advantage of OVL's portability will diminish. On the other hand, OVL has new opportunities to expand its list of hosting languages. In fact, OVL [Acc03a] assertions already have embedded PSL pragmas.

One drawback inherent in the collection-of-assertions approach, however, is that you only get what is offered in the library. One possible remedy to this limitation is to make the assertions composable. This can be accomplished by returning, from each assertion, a status (true/false/pending) that may then be used as a condition by other assertions.

EXAMPLE 4.1   The OVL "assert_next" may be modified to compose a sequence as follows. Add one output called "status" to the definition of "assert_next" so that the declaration looks like:

```
assert_next name (clk, reset_n, start, test, status);
```

where "status" is

- X if "start" has not been asserted, or the time specified by "cycles" has not yet expired.

- 0 if "start" has been asserted, the time has expired, and "test" is not asserted.

- 1 if "start" and "test" are asserted.

Then, the following Verilog code specifies the assertion for the sequence of events "if I, then next A followed by B followed by C":

```
wire statA, statB, statC;
assert_next step1 (clk, reset_n, I, A, statA);
assert_next step2 (clk, reset_n, statA, B, statB);
assert_next step3 (clk, reset_n, statB, C, statC);
assert_never (clk, reset_n, statA==0 || statB ==0 || statC==0);
```

□

The above example is similar to the *endpoint* construct in PSL and SVA. Another possibility for extending OVL is to allow recursive definitions, as in SVA.

4.2   Temporal Logic and Regular Expression

In the sequel, we denote logic conjunction, disjunction, implication, and negation operators with ∧, ∨, →, and !, respectively. Be careful not to confuse negation with PSL's strong operator, !, which is left-associative (after the operand).

4.2.1 Temporal Logic

Temporal logic describes and allows inference about behaviors over time. The two types of temporal logics that are used the most are the linear temporal logic LTL and the Computational Tree Logic (CTL). The building blocks of both logics are the *Atomic Propositions* (AP), which, in the case of design verification, correspond to design signals.

Propositions are composed from AP with Boolean connectives such as conjunction, disjunction, and negation. LTL reasons about the truthness of propositions along a linear discrete timeline, whereas CTL does so along a tree-like timeline where, at each moment in time, there may be multiple futures. LTL formulas are inductively defined as follows.

DEFINITION 4.1 (LTL) All APs are LTL formulas; if p and q are LTL formulas, so are the following:

- $!p$ - $p \wedge q$ - $p \vee q$ - Xp - $p \, U \, q$ - $p \, W \, q$.

At a given state, an AP is either true or false based on the value of the corresponding signal. Inductively, Xp is true iff (if and only if) p is true in the next time step; $p \, U \, q$ is true iff p is true until q is true and q must be true eventually; and pWq is true iff p is true as long as q is not true, and unlike U (*strong* until), q does not have to be true in the future.

There are also shorthand ways of expressing commonly used formulas: Fq (*finally* or *eventually* q) stands for *true* $U \, q$, and Gp (*always* p) stands for $!F!q$.

Many useful properties can be readily expressed in LTL. For example, the property about a "fair" bus arbiter

Whenever there is a request, there will be a grant

can be written as the LTL

$$G(request \rightarrow F(grant)). \tag{4.1}$$

CTL adds branching operators to LTL. A CTL formula is either an AP, or a nested formula with alternating branching operators (E as *for some* and A as *for all*) and temporal operators. The outermost operator must be a branching operator. CTL formulas are defined mutually recursively with "path formulas", as in the following.

DEFINITION 4.2 (CTL) If p and q are CTL formulas, then the following are path formulas:

- Xp - $p \, U \, q$ - $p \, W \, q$.

All APs are CTL formulas. If p and q are CTL formulas, and r is a path formula, then the following are CTL formulas:

- $!p$ • $p \wedge q$ • $p \vee q$ • $E\,r$ • $A\,r$.

The addition of branching operators enables CTL to express properties that are not possible in LTL. For example, the "dead-lock free" property

> From any state p, there is a way to get out of p

can be written as the CTL

$$AG(p \rightarrow EF!p). \tag{4.2}$$

On the other hand, some properties are expressible in LTL but not in CTL due to the restriction that branching and temporal operators must interleave in a CTL formula. The fairness property in Formula (4.1) is such an example, because the temporal operators G and F appear consecutively. Another useful property

> On every execution path there is a state after which p is always true.

is an LTL formula with G and F reversed, as below.

$$FG(reset\_done).$$

This is not expressible in CTL either.

An extension to CTL, called CTL, removes the alternating restriction and only requires that the outermost operator be a branching operator. CTL* subsumes CTL and LTL in expressiveness.[1] The disjunction of the CTL formula (4.2) and the LTL formula (4.2.1) is a CTL* formula that is neither CTL nor LTL:

$$AG(p \rightarrow EF!p) \vee FG(reset\_done).$$

4.2.2   Regular Expression

There are still significant shortcomings in what can be expressed with the temporal languages just introduced. A famous example is the property that "at all even times $(0, 2, 4, \ldots)$, q is true" [Wol83]. However, this can be easily captured by the *regular expression* below:

$$q \cdot (true \cdot q) * .$$

---

[1] Conventionally, an LTL formula p is interpreted as Ap.

Regular expressions (REs) define finite sequences of letters (these sequences are also called *strings* or *words*). They are widely used for pattern matching, for example, in UNIX shell commands and PERL built-in functions. Let $\sum$ be the alphabet of letters, ε a special *empty* letter, and $\emptyset$ the empty set that contains no words. Then RE is defined by the following.

DEFINITION 4.3 (REGULAR EXPRESSION) Any letter from $\sum$, ε, and $\emptyset$, are REs. If p and q are REs, so are the following:

$$\bullet \, p \cdot q \qquad \bullet \, p \mid q \qquad \bullet \, q * .$$

The set of words corresponding to an RE is obtained as follows. The RE of a single letter a gives the set $\{a\}$, ε gives the set containing the empty letter $\{\varepsilon\}$, and $\emptyset$ gives the empty set $\{\}$. The RE $p \cdot q$ gives the set of *concatenations* of strings in the sets of p and q, while $p \mid q$ gives the *union* of the two sets. Finally, $p*$, where "$*$" is called the *Kleene star* operator represents the set of zero or more concatenations of strings in the set of p. Zero concatenation results in the empty letter ε. The sets of words derived in this way from REs are sometimes referred to as *regular languages*.

EXAMPLE 4.2 Let $\sum = \{a, b, c\}$. Below is a list of REs and their corresponding set.

$$
\begin{aligned}
a \cdot b &= \{ab\} \\
a \mid b &= \{a, b\} \\
(a \cdot b) \cdot (a \mid b) &= \{aba, abb\} \\
(a \cdot b) \mid (a \mid b) &= \{a, b, ab\} \\
(a \cdot b)* &= \{\varepsilon, ab, abab, ababab, \ldots\}
\end{aligned}
$$

$\square$

4.2.3 Truthness of Properties

Temporal logics and regular expressions specify an unambiguous set of behaviors. Design verification with respect to a property in temporal logic is essentially the process of deciding if the design behavior conforms to (i.e., is a subset of) the behavior specified by the property. LTL and regular expressions model design behavior as linear execution paths. Whether or not a design execution belongs to these paths can be checked in simulation, as well as in formal verification, whereas CTL models design behavior as execution trees, which can only be checked in formal verification. We elaborate more on the truthness of LTL and regular expression formulas along a path.

LTL, as defined earlier, uses the *post-time* modality, that is, time advances forward. For example, the X and U operators refer to the modalities "next" and "finally," respectively. Furthermore, "finally" can be inductively defined by "next." For example, $p\ U\ q$ means:

$$q \vee p \wedge X(p\ U\ q).$$

Since all LTLs that are not propositions involve X or U, they can be rewritten using X. A less obvious example, "p is always true" $(G(p))$, is equivalent to

$$p \wedge X(G(p)).$$

Similarly, regular expressions can also be rewritten by borrowing the X operator. For example, $a \cdot b$ is rewritten as

$$a \wedge X(b),$$

and the regular expression for "p is true in the future," $(true*) \cdot p$, is rewritten as

$$p \vee X((true*) \cdot p).$$

The expressions that follow the X operator can be understood as the "future obligation," and the others as the "current obligation." For a property to hold, all current obligations must be satisfied, and if they are not, we say we have encountered a bad state. The future obligations are guarded by the current ones, so they may or may not be applicable. In the first example, the current obligation is "q or p". If q is satisfied, then the future obligation $X(p\ U\ q)$ is discharged. In the second example, however, the current and future obligations p and $X(G(p))$ must be satisfied at all times.

The above analysis provides a basis for interpreting the truth of a property on a path at any given time. Therefore, it is suitable for finite, truncated paths as seen in simulation or in bounded model checking. Specifically, there are three situations:

1. Property holds:

    a. no bad state has been encountered, and

    b. all future obligations are discharged

2. Property is pending:

    a. no bad state has been encountered, and

b. some future obligations are not discharged

3. Property fails:

some bad state has been encountered

4.2.4 Strong and Weak Semantics

In the previous section, the future obligation of a property is withheld regardless of whether or not there is a future about the path. This is called "strong semantics." In "weak semantics," the future obligations do not have to be satisfied for the property to hold; on a truncated path, a property holds as long as there is not bad state.

Strong and weak semantics can co-exist if we use different operators for the two versions of a temporal operation and modify the definition of property satisfaction.[2] For example, we can use the "weak_until" (pWq) as opposed to (strong) "until" ($p\ U\ q$). The formula pWq is weaker than $p\ U\ q$ in that the first formula is true if either the second formula is true or q in the first formula never occurs. In the latter case, p must hold forever. This is captured in the following identity.

$$pWq = (p\ U\ q) \vee G(p).$$

Incidentally, weak_until requires a weak_next (X_w), such that

$$pWq \leftrightarrow q \vee p \wedge X_w(pWq).$$

where $X_w(p)$ is true if either p is true next or the trace terminates before next.

Then, if we trace the satisfiability of future obligations from weak operators, we arrive at a stronger version of hold in which future obligations from both strong and weak operators are discharged. The property is said to "hold strongly" in this case. As a result, we add one more situation to have a 4-state status regarding property satisfaction on a truncated path:

1. Property holds strongly:

    a. no bad state has been encountered, and

    b. all strong and weak future obligations are discharged

2. Property holds:

    a. no bad state has been encountered, and

    b. all strong future obligations are discharged

---

[2]There are no weak operators in regular expressions. Instead, PSL uses $r!$ to denote the weak version of the regular expression r.

3. Property is pending:

   a. no bad state has been encountered, and

   b. some strong future obligations are not discharged

4. Property fails:

   some bad state has been encountered

The co-existence of strong and weak operators can be understood as the artifact of a language, in this case LTL, being "closed" under negation. That is, for any LTL p, the formula $!p$ is also an LTL. Instead of carrying $!$ in front of all negated LTLs, we can push the negation down to the (atomic) propositions by negating the temporal operators and applying DeMorgan's Law to the Boolean connectives. Because negation happens only at the lowest possible level, an LTL formula obtained in this way is said to be in a Positive Normal Form (PNF). However, in order to push the negation through the temporal operators, we must define the negated forms, or the duals, of these operators.

It turns out that the dual of a strong temporal operator is a weak operator, and vice versa. The dual of X is the the weak_next operator X_w previously introduced to define weak_until. Therefore, we have

$$!X(p) = X_w(!p).$$

The duals of the strong and weak untils are the weak and strong *release* operators, respectively. The weak_release operator, R, is defined as

$$p \, R \, q \leftrightarrow q \wedge (p \vee X_w(p \, R \, q))$$

and we have

$$!(p \, U \, q) = !p \, R \, !q.$$

The definition of strong_release and its duality with respect to W are similarly to the above.

Previously, we shorthanded $!F(!p)$ with $G(p)$. With weak_release, we can define $G(p)$ as $(false \, R \, p)$. The duality between (strong) F and (weak) G is also obvious.

With these extensions, LTL is closed under negation. To apply the previous 4-state definition of satisfiability, an LTL formula is first put into PNF so that no operator is inside the scope of a negation. Then, the strong and weak obligations can be clearly identified. Note, however, that this succinctness of LTL negation due to the duality of strong and weak operators does not occur in Regular Expression. For example, the negation of $a \cdot b \cdot c$ would give

$$(\varepsilon) \mid (!a \cdot T*) \mid (a \cdot !b \cdot T*) \mid (a \cdot b \cdot !c \cdot T*) \mid (a \cdot b \cdot c \cdot T \cdot T*).$$

Therefore, in PSL, the negation and strength of Regular Expression are expressed explicitly outside of the expression. For example, the negation of $a \cdot b \cdot c$ is simply $!\{a \cdot b \cdot c\}$, and the strong version is $\{a \cdot b \cdot c\}!$. The expression by itself is understood as being weak.

4.2.5 Safety and Liveness Properties

Strong operator U gives rise to future obligations that must be satisfied. Meanwhile, the fulfillment of these obligations, called *eventualities*, are not bounded in time. Whether or not the obligations are bounded leads to the frequently used *safety* and *liveness* classification of properties.

Intuitively, safety means that something bad never happens, whereas liveness means something good can eventually happen. While there are various slightly different versions, the following definitions about safety and liveness are more common:

DEFINITION 4.4 (SAFETY) A property is a *Safety* property if every failing path has a finite prefix: the prefix, or any of its extensions, is also a failing path.

DEFINITION 4.5 (LIVENESS) A property is a *Liveness* property if any finite path can be extended to satisfy that property.

It follows that a safety property cannot simultaneously be a liveness property (except for the tautology "true," which does not fail on any path). To see this, take any safety property, and a path on which it fails. Find the prefix of the path. Therefore, any extension of the prefix must also fail the property. That is, the prefix cannot be extended to satisfy the property. Hence the property cannot be a liveness property.

Some properties are neither safety nor liveness. For example, $p\ U\ q$, although requiring that q eventually be true, falls into this category. It fails on an infinite path where q never holds and this path has no finite prefix on which the property fails; meanwhile, the path (of length 1) $!p!q$ cannot be extended in any way to satisfy the property.

Yet another type of property is the so-called *co-safety* property. A property is co-safety if its negation is a safety property. Therefore, by replacing "failing" with "satisfying" in the definition of safety, we get the definition of co-safety.

DEFINITION 4.6 (CO-SAFETY) A co-safety property is a property where every satisfying path has a finite prefix: the prefix, or any of its extensions, is also a satisfying path.

What is interesting about co-safety is that it overlaps with liveness: the co-safety property $!G(p)$ is exactly the liveness property $F(!p)$. Co-safety also overlaps with safety in, for example, the propositional property p. There are

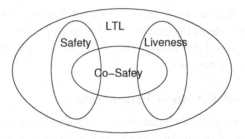

Figure 4.1. Safety, liveness, and co-safety.

also co-safety properties that are neither safety nor liveness, like the property $p \, U \, q$. The relation among these characterizations is illustrated in Figure 4.1.

The main motivation for these characterizations is that most industrial properties are safety properties, and there are specialized algorithms that handle safety more efficiently than the generic algorithms. Specifically,

- In *static verification*, verification of safety properties reduces to a reachability problem, whereas verification of non-safety properties in general requires the additional detection of loops. Loop detection can be just as complex as the reachability computation.

- In *dynamic verification and constrained random generation*, checkers and generators for safety or co-safety properties (constraints) are simple finite state automata. In general, the more costly (Büchi [Büc60]) automata are required. We will elaborate on property checkers and constraint generators later in this chapter.

- The class of properties that are both safety and co-safety, called *bounded specifications* [KV01], can be proven and falsified in a bounded number of cycles, as evident in the definitions of safety and co-safety. The bound is determined by the specification itself [KV01]. With a known bound, bounded model checking and simulation become both sound and complete.

A complete procedure to decide if an LTL property is a safety or co-safety property is hard (PSPACE-complete [KV99]). However, there is a simple syntactical rule that infers safety and co-safety. We consider LTL formulas in PNF. An LTL formula whose only temporal operators are X, X_w, and R is syntactically safe. Similarly, an LTL formula whose only temporal operators are X, X_w, and U is syntactically co-safe. It follows that any LTL formula in which the only temporal operators are X and X_w is both safe and co-safe, i.e., bounded.

4.2.6 Multiple Paths and Initial States

It is almost certain that designs have more than one execution path. As LTL and regular expressions are defined over a single path, generally the property is checked with the "for all path" quantifier. Therefore, simulation may discover failures but never declare a pass, even if the property holds on all paths simulated.

There are situations where the "for some path" quantification is useful, such as the negation between safety and co-safety described. Another example is that to prove a state s is not deadlocked, one may use s as the initial state and prove that "there is an execution such that $F(!s)$." In this case, simulation can actually prove the property holds but cannot prove the reverse. In any case, there seems to be sufficient evidence that both path quantifiers should be handled, and the simulation results based on the single-path semantics interpreted accordingly. Note that this poses no extra requirement for CTL.

By the same argument, the distinction between universal and existential quantifications also applies to initial states. When there are multiple initial states, proving a property requires the examination of all initial states.

4.3 Introduction to PSL

By introducing temporal logics, regular expressions, the truthness of properties, and strong and weak semantics, we have covered the most arcane concepts of the PSL language. Specifically, these are the main constituents of the *temporal layer* of PSL. In this section, we intend to give an overview of the PSL language constructs, illustrating what we have already discussed. For detailed syntax and semantics, refer to the PSL-1.1 Language Reference Manual (LRM) [Acc04].

4.3.1 The Four Layers of PSL

A PSL property consists of four layers: *Boolean, temporal, verification,* and *modeling*.

- **Boolean layer**

  This layer specifies propositions, or expressions over design and auxiliary signals, that evaluate to true or false in a single evaluation cycle. The expressions are written in the HDL that describes the design. Four flavors of languages are allowed: Verilog, VHDL, SystemVerilog, and GDL (General Description Language).

- **Temporal layer**

  This layer is the heart of the language, and is used to describe the temporal behaviors built up with Boolean layer propositions and temporal operators. The temporal layer consists of the Foundation Language (FL) and the Optional Branching Extension (OBE). FL is a linear temporal logic that embeds

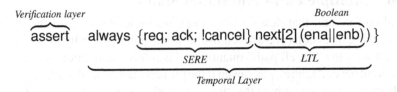

Figure 4.2.   A PSL assertion with 3 layers.

within LTL a customized version of Regular Expression, called the Sequential Extended Regular Expression (SERE). OBE is the CTL language. FL and OBE cannot be mixed in one property.

- **Verification layer**

  This layer gives the verification directives, that is, explains what to do with the properties described by the temporal layer. There are seven directives: assert, assume, assume_guarantee, restrict, restrict_guarantee, cover, and fairness.

- **Modeling layer**

  This layer is used to model the input behavior (for formal verification), and adds auxiliary signals and logic that are needed for verification. Like the Boolean layers, the modeling layer also comes in four flavors: Verilog, VHDL, SystemVerilog, and GDL. This layer is optional.

Figure 4.2 demonstrates a PSL assertion with three layers.

4.3.2   Verification Units

PSL statements can either be stand-alone, as in the previous example, or grouped into *verification units*. There are three types of verification units: *vprop*, which groups assertions to be verified, *vmode*, which groups constraints (with the assume/restrict directives), and *vunit*, which combines the two. Because constraints specify necessary conditions for proving the assertions, the two are usually separated. Verification units may also contain modeling layer constructs used by the assertions or constraints. The following is an example.

```
vunit hand_shake (top.bus) {
    wire temp;      // modeling layer
    assign temp = ack1 || ack2;      // modeling layer
    assert always (reqa -> next temp);
}
```

A verification unit can also inherit other verification units by using the *inherit* statement. For example:

```
vmode consB (top.blockB) {
    inherit Common;      // "Common" was defined before
    assume never (grantB1 && grantB2);
}
```

Although not required for PSL assertions and constraints, verification units provide basic OOP capabilities that facilitate a modularized approach.

4.3.3   Sequential Extended Regular Expression

Sequential Extended Regular Expression (SERE) is an extension to the Regular Expression introduced earlier. Table 4.2 gives the syntax and descriptions of SERE.

Table 4.2.   Definition of SERE.

SERE	Description
[*]	skip 0 or more cycles
[+]	skip 1 or more cycles
[*n]	skip n cycles
<Boolean>	Boolean (expression) is true immediately
<SERE>[*]	SERE repeats 0 or more times consecutively
<SERE>[+]	SERE repeats 1 or more times consecutively
<SERE>[*n]	SERE repeats n times consecutively
<SERE>[*n:m]	SERE repeats between n and m times consecutively
<Boolean>[=n]	Boolean repeats n times
<Boolean>[=n:m]	Boolean repeats between n and m times
<Boolean>[->n]	Boolean repeats n times, the last occurs at the end of the path
<Boolean>[->n:m]	Boolean repeats between n and m times, the last occurs at the end of the path
{SERE}	Braced SERE
{SERE} @ <clock>	Clocked SERE
<SERE> ; <SERE>	Concatenation
<SERE> : <SERE>	Fusion − the second starts at the last cycle of the first
<SERE> I <SERE>	Union
<SERE> & <SERE>	Intersection
<SERE> && <SERE>	Intersection with length matching
<SERE> within <SERE>	The second contains the first

4.3.3.1   Clocked SEREs

Clocked SERE is not a straightforward extension to regular expression. For a clocked SERE to be satisfied, the clock must evaluate to 1 at all steps in the SERE. The clock is level-sensitive if it is a Boolean expression; an edge sensitive clock can be expressed with built-in functions, as in $\{SERE\}@(rose\ clock)$

or $\{SERE\}@(\text{fell clock})$. There can also be nested clocks, where the inner clocks takes precedence over the outer ones. For example, $\{a; b@clk1; c\}@clk2$ is equivalent to $\{a@clk2; b@clk1; c@clk2\}$.

It should be noted that every clocked SERE can be translated to an equivalent unclocked SERE. A set of rewriting rules is given in Appendix B of PSL-v1.1 LRM [Acc04].

4.3.3.2 Sequences

SEREs can be used as a stand-alone property or embedded into an LTL property, both in the form of *sequences*. A sequence is simply a braced or clocked SERE. Negation and strong operator, which is not available for SEREs, are applicable to sequences. To negate or obtain the strong version of a sequence, simply prepend or append it, respectively, with !, as in !$\{a; b\}$ and $\{a; b\}$!.

4.3.3.3 SEREs on Infinite Paths

The definition of SERE given in Table 4.2 applies to paths of finite length, as is also the case with regular expression. Satisfaction of a SERE on a finite path requires an exact match. For example, $a; b$ is satisfied by the path of length 2: a is true then b is true. Any path shorter or longer than 2 would not be a match. This method of satisfaction is referred to as the SERE "holds tightly" on the finite path.

The satisfaction of sequences, the embedding of SEREs in LTL, is defined by using "holds tightly." With the addition of strong semantics to sequences, whether or not a sequence holds on a finite or infinite path is defined as follows:

- A weak sequence (without ! at the end) holds on a path iff the corresponding SERE holds tightly on an extension or on a prefix of the path.

- A strong sequence (with ! at the end) holds on a path iff the corresponding SERE holds tightly on a prefix of the path.

As a result, the (weak) sequence $\{a; b\}$ holds on the one-cycle path "a is true," while the strong sequence $\{a; b\}$! does not hold on this path, but does hold any path of 2 or more cycles where the first two cycles are "a is true, then b is true."

4.3.4   The Foundation Language

The Foundation Language of PSL is LTL extended with SERE, as given in Table 4.3. The referred unary and binary temporal operators are described in Tables 4.4 and 4.5.

Note that in Tables 4.4 and 4.5, except for F (or **eventually!**) and **abort**, all the other FLs are weak operators. As we recall, a property defined with

Table 4.3. Definition of FL.

FL	Description
<Boolean>	Boolean
<Sequence>	Sequence
<Sequence>!	Strong sequence
NOT_OP <FL>	Negation with language flavor
<FL1> BIN_OP <FL2>	Boolean connectives with language flavor
<FL1> -> <FL1>	FL1 implies FL2
<FL1> <-> <FL2>	FL1 and FL2 are equivalent
UNA_TEMP_OP <FL>	Unary temporal of FL
<FL1> BIN_TEMP_OP <FL2>	Binary temporal of FL1 and FL2
<Sequence> |-> <FL>, <Sequence>(<FL>)	If Sequence holds, then FL holds in the last cycle of the prefix on which Sequence holds tightly
<Sequence> |=> <FL>	If Sequence holds, then FL holds in the cycle right after the prefix on which Sequence holds tightly

Table 4.4. Definition of weak unary temporal operators.

FL	Description
F <FL>, **eventually!** <FL>	FL holds eventually
G <FL>, **always** <FL>	FL always holds
X <FL>, **next** <FL>	FL holds in the next cycle
X [n]<FL>, **next** [n]<FL>	FL holds n cycles later
next_a [n:m]<FL>	FL holds in all the cycles in the range
next_e [n:m]<FL>	FL holds in at least one cycle in the range
next_event (<Boolean>)<FL>	FL holds at the next occurrence of Boolean
next_event (<Boolean>)[n]<FL>	FL holds at the n'th occurrence of Boolean
next_event_a (<Boolean>)[n:m]<FL>	FL holds always in the range of occurrences of Boolean
next_event_e (<Boolean>)[n:m]<FL>	FL holds at least once in the range of occurrences of Boolean

a weak operator needs to hold only if the computation path is not truncated inopportunely before the expected cycles or events can happen. For example, if simulation terminates right after the current cycle, that is, there is no next cycle, then $X(p)$ is released of the obligation that p holds next. Therefore, the property holds (but not strongly). Similarly, for property $next\_event(p)[5](q)$, if the simulation stops before the 5th occurrence of p, then the property also holds.

Table 4.5.   Definition of weak binary temporal operators.

FL	Description
<FL1> **until** <FL2>	FL1 holds up to the cycle FL2 holds
<FL1> **until_** <FL2>	FL1 holds up to and including the cycle FL2 holds
<FL1> **before** <FL2>	FL1 holds before FL2 holds
<FL1> **before_** <FL2>	FL1 holds before or at the same cycle as FL2 holds
<FL1> **abort** <Boolean>	FL1 holds unless Boolean evaluates to true first

Each of these weak operators, except for G, whose negation is the strong operator F, has a strong version with the indicator ! appended to its key word. For example, the strong version of $X(p)$ and $next\_event(p)[5](q)$ are $X!(p)$ and $next\_event!(p)[5](q)$, respectively. The strong operators demand that the property hold unconditionally. In the above examples, the strong versions of the properties would fail in the truncated simulation path.

4.4   Monitors and Generators

Now that we have given an overview of the PSL language, we will continue by showing how the language may be used, or implemented, with verification tools. Since this book is about constrained simulation, we will be interested mainly in how PSL may be applied in simulation. Therefore, OBE, the branching time part of PSL that is suitable only for static verification, will be avoided. We refer interested readers to *CTL model checking* [CE81, EL85, McM93, BHSV⁺96] for further study of OBE.

This is not to say that the whole FL part of PSL is suitable for simulation. In this section we will show that many FL properties are *simulatable* in the sense that they can function as monitors or generators. Monitor and generator are in fact the dual usages of a PSL property, corresponding to the *assertion* and *assume/restrict* verification directives.[3]

4.4.1   Monitors

A monitor surveys the state of the design during simulation, upon which it deduces the status of the property it represents. In testbench, monitored properties are the functional coverage points. Monitors are finite state machines that accept or reject certain simulation traces. Some states in the monitor are

---

[3]The PSL LRM does not mention the use of PSL properties as constraints in testbench, but neither does it exclude testbench when defining that the assume directive "instructs the verification tool to constrain the verification (e.g., the behavior of the input signals) so that a property holds".

designated as *accepting* or *rejecting* states. Roughly speaking, a simulation trace that drives the monitor into an accepting (rejecting) state exhibits a good (bad) behavior.

Exactly how monitors work goes back to the so-called "language-theoretic" approach to the analysis of formal languages. Historically, regular expressions and temporal logics are characterized as language recognizers. The language here refers to a set of words over a finite collection of alphabets. The words recognized by a regular expression have finite length, in pattern matching, for example, and the words recognized by an LTL have infinite length, useful in modeling design executions. Whether or not a word belongs to a language can be procedurally determined by running the word over a Nondeterministic Finite Automaton (NFA) in the case of RE, or a nondeterministic Büchi automaton in the case of LTL, as in the following:

1. **The automaton**: Some states from the automaton are marked as the initial states, and some as accepting states; the automaton's transitions are labeled with letters from the alphabet, and a state may have many outgoing transitions labeled with the same letter.

2. **The run of a word**: A word runs over the automaton by starting from all initial states, following the transitions corresponding to the sequence of letters in the word. In general, there can be many paths of such transitions corresponding to the run of one word. This is because 1) there can be multiple initial states, and 2) each state can transition to multiple successors on the same letter.

3. **Accepting a word**: In the case of a regular expression, if the final state on any of the paths is also an accepting state, then the word is accepted. For LTL, a word is accepted if any of the paths visits an accepting state infinitely often (that is, there is a loop on the path that contains an accepting state).

4. **Rejecting a word**: If a word is not accepted, then it is rejected. However, we should distinguish between these two scenarios: 1) the run has finished (only for NFA) but is not accepted; 2) the run is terminated because the last state has no exiting transition whose label matches the current letter.

EXAMPLE 4.3   Figure 4.3 gives the NFA of the regular expression $(a \cdot b)*$, and Figure 4.4 gives the Büchi automaton of the LTL formula $p \ U \ q$. The short diagonal arrows point to the initial states. States with a double-circle are the accepting states.

The language accepted by the NFA is $\{\varepsilon, ab, abab, ababab, \ldots\}$. The empty letter ε is accepted because the accepting state (labeled $s3$) is also an initial state. To see why the other words are accepted, we start from the other initial

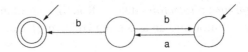

Figure 4.3.   NFA for $(a \cdot b)*$.

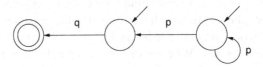

Figure 4.4.   NFA for $p \ U \ q$.

state, $s1$. The first letter has to be a because the only exiting transition is labeled
a — any letter other than a would terminate the run and lead to an immediate
rejection. Similarly, the second letter has to be b. The letter b enables two
transitions, one to $s3$, which accepts the word ab, while the other goes back to
$s1$ and the process repeats, as shown below.

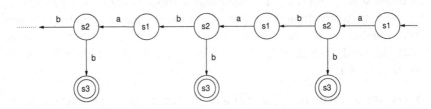

Figure 4.5.   The runs for $(a \cdot b)*$.

The language accepted by the Büchi automaton, if written as regular expressions, is $\{q, pq, ppq, pppq, \ldots\}$, where the sequence of p's can be infinite. Note
that the "visiting the accepting state infinitely often" part is not enforced in the
Büchi automaton. Rather, it is implemented by the underlying verification tool.
□

There are, however, two problems in applying the above flow to a real simulation: one, it is hard to trace multiple paths, and two, testing the Büchi
automaton accepting condition for LTL is cumbersome (which is true even for
static verification).

To address the first problem, the automaton needs to be *determinized* so that
there is one initial state, and that a state transitions to at most one successor
on any letter. For NFA, a simple method called *subset construction* can be

used. However, determinizing Büchi automata is much more involved [McN66, Saf88].

Fortunately, Büchi automata are avoided altogether in our attempt to address the second problem. Recall that for safety (co-safety) properties, every failing (satisfying) path has a bad (good) prefix, that "tells the whole story" [KV99] about the failure or success of an LTL property. Since the prefixes are finite, an NFA suffices to recognize them.[4]

Most industrial properties are safety and co-safety properties. Note that, as shown in Figure 4.1, co-safety properties overlap with liveness properties. Therefore, prefix-detection-based NFA can monitor a large subset of properties. We will give details on constructing such NFAs and on how they are determinized in Section 4.5.

4.4.2 Generators

In the previous monitor example, the runs of the RE $(a \cdot b)*$ in Figure 4.5 have already given an idea of how a monitor may be used as a generator: suppose a and b are conditions that involve inputs to the design, and following the transitions in the automaton imposes certain behavior on the inputs that conforms to the conditions. That is, inputs are generated.

In other words, when monitoring a property, which transition to take in the automaton is immediately known because the letters are determined only by the design state. In generators, this is known only after conditions involving inputs are solved and a solution to the inputs has been picked.[5] More specifically, a generator works as in the following:

- Take the disjunction (OR) of the propositions corresponding to the letters of all transitions out of the current automaton state;

- Plug the current state assignment in to the disjunction and solve for the inputs;

- If there is no solution, then the constraint is violated;

- Otherwise, randomly choose a solution and follow the transition whose label (a proposition) evaluates to true under this solution.

Note that the generator automaton has to be deterministic so that each solution to the inputs makes exactly one transition's label true.

---

[4]The *pathological* safety properties have no bad prefixes, and so cannot be recognized by an NFA [KV99]. However, these properties are highly redundant, a condition rarely encountered in reality. Additionally, they can be reduced to equivalent *natural* safety properties that can be recognized by NFAs.

[5]It is very unusual for a property to depend on any input, because it is always possible to falsify the property. Also, constraints that do not depend on any input are very special; they are treated as properties to be checked instead of as input-generating constraints.

4.5 Monitor Construction

The construction of language-recognizing automata for REs and LTLs has a long history. The earliest RE to automata translations were given in [MY60, Tho68]. LTL to automata translation was pioneered in [WVS83], which is *tableau*-based. In a *tableau-based approach*, the satisfaction of a temporal formula is decomposed both logically (across Boolean connectives) and temporally (obligations in the next-time), as explained in Section 4.2.3.

Due to the increasing interest in the static and dynamic verification of temporal properties, there is a range of recent studies on how to efficiently convert LTLs to automata [DGV99, SB00, GO01, GL02, Thi02]. Since the standardization of PSL, there are also attempts to convert PSL to automata, for example, [GHS03, CM04, GG05, BFH05]. Some give the conversion of a subset of PSL. Additionally, all EDA tools that support PSL (for example, *FoCs* from IBM [ABG$^+$00]) must perform similar conversions. A complete treatment of PSL to automata conversion is quite complicated. Some of these complications are listed below.

- The combination of RE (of finite words) and LTL (of infinite words).

- The strong and weak semantics on truncated paths.

- New operators such as fusion and sequence implication.

- Negation of RE.

Take the negation of RE as an example. Unlike LTL formulas, there is no straightforward and closed-form negation of REs. Instead, to negate an RE, one builds the NFA of the RE, converts it to a DFA, complements the DFA, and finally constructs the negated RE from the resulting DFA. Therefore, in general, tableau-based approaches are not suitable for negated REs.

In this section, we select a subset of FL in PSL and demonstrate the construction of finite automata to be used as monitors or generators. The translation is tableau-based, but with an extension to handle (positive) REs and their embedding in LTL. The subset of FL, referred to as FL$^-$, is defined below.

DEFINITION 4.7 (FL$^-$) All Boolean formulas are FL$^-$ formulas; Let r be an RE defined in Definition 4.3, and let p and q be FL$^-$ formulas, then the following are also FL$^-$ formulas:

$$\bullet\, r \quad \bullet\, p \wedge q \quad \bullet\, p \vee q \quad \bullet\, Xp \quad \bullet\, p\,U\,q \quad \bullet\, p\,R\,q.$$

Therefore, all FL$^-$ formulas are in PNF. Note that the exclusion of negation does not restrict the expressiveness of pure LTL formulas that have no RE embedding. The shorthands $F(p)$ and $G(p)$ are also FL$^-$ formulas. An FL$^-$

formula is first converted to an NFA using a set of tableau rules, then the NFA is converted to a DFA using the standard subset construction approach. The DFA is now ready to be used as a monitor or generator in simulation.

4.5.1 The Tableau Rules and Cover

Tableau rules decompose the satisfiability of an FL⁻ formula into a disjunction of obligations. Each obligation may be a proposition, a formula to be satisfied next, or the conjunction of both. The proposition and the formula means the current and next-time sub-obligations, respectively. This disjunctive decomposition provides choices of how the property may be satisfied, signifying the nondeterminism in the automaton to be built. The current/next-time partitioning enforces the temporality.

EXAMPLE 4.4 To satisfy the LTL $p\,U\,q$, we either need to satisfy q at the current time and nothing next, or p at the current time and $p\,U\,q$ at the next time. The first disjunct releases all obligations, and hence by itself satisfies the property. The second disjunct makes sure that p holds currently and that the original property needs to hold again in the next time since it is not yet fulfilled. □

The disjunctive obligations of a formula are called the *cover* of the formula. To see how the cover is computed, we first introduce the notion $N(p)$ to mean that p is the next-time obligation. Note that $N(p) \land N(q) = N(p \land q)$ and $!N(p) = N(!p)$.

Although the cover computation for the LTL and RE parts of an FL⁻ formula can occur simultaneously, we will separate the discussion for clarity. For an LTL formula, its cover is computed as follows:

1. Apply the following rewriting rules until the result consists of only propositions and subformulas of the form $N(p)$:

$$
\begin{aligned}
p\,U\,q &= p \lor (p \land N(p\,U\,q)) \\
p\,R\,q &= p \land (p \lor N(p\,R\,q)) \\
F(p) &= p \lor N(F(p)) \\
G(p) &= p \land N(G(p))
\end{aligned}
$$

2. Put the result in disjunctive normal form.

Computing the cover of an RE is somewhat involved. The main idea is to derive the disjunctive normal form:

$$s_1 \mid s_2 \mid \ldots \mid s_n,$$

where s_i is either the empty letter ε, a proposition, or an RE in the form $b \cdot r$ where b is a proposition and r is an RE. This way, the current and next-time obligations become obvious. The computation proceeds as follows:

1. The distribution rules below are applied to the RE until they are not applicable:

$$(r \mid s) \cdot t = r \cdot t \mid s \cdot t$$
$$t \cdot (r \mid s) = t \cdot r \mid t \cdot s$$

Note that $\varepsilon \cdot r = r \cdot \varepsilon = r$. The result is a union ($|$) of REs where in each constituent RE the operator $|$ only occurs in the operant of $*$.

2. If there are constituent REs of the form $r*$ or $r* \cdot s$, then apply the following $*$-rewriting rule and go back to step 1.

$$r* = r \cdot r* \mid \varepsilon.$$

3. Finally, replace each top-level $|$ with $\vee$, and for each constituent RE of the form $b \cdot r$, where b is a proposition and r is an RE, apply

$$b \cdot r = b \wedge N(r).$$

EXAMPLE 4.5   Normalization of $(a* \cdot b)*$ proceeds as follows.

$$\begin{aligned}
(a* \cdot b)* &= (a* \cdot b) \cdot (a* \cdot b)* \mid \varepsilon \\
&= ((a \cdot a* \mid \varepsilon) \cdot b) \cdot (a* \cdot b)* \mid \varepsilon \\
&= ((a \cdot a* \cdot b \mid b) \cdot (a* \cdot b)* \mid \varepsilon \\
&= ((a \cdot a* \cdot b \cdot (a* \cdot b)* \mid b \cdot (a* \cdot b)* \mid \varepsilon
\end{aligned}$$

Therefore, the cover is

$$a \wedge N(a* \cdot b \cdot (a* \cdot b)*) \ \vee \ b \wedge N((a* \cdot b)*) \ \vee \ \varepsilon.$$

$\square$

4.5.2   Constructing the NFA

Construction of the NFA for an FL^- is based on the repeated computation of covers. The cover of an FL^- formula consists of a set of current and next-time obligations. The former can be asserted immediately; the latter is carried over to the next-time for which we compute the cover (if it is not already computed) which may again generate more next-time obligations. This process eventually terminates because the previously given rewriting rules only generate subformulas of the original formula, which are finite.

The covers computed this way represent all the states in the NFA: each unique disjunct, or *term*, in a cover corresponds to a state. The term labels the state. The NFA can then be constructed as below.

1. Let the initial states be the ones in the cover of the original formula.

2. Add a transition from state A to state B if B contains a subformula that is the next-time obligation of a subformula in A.

3. Add an additional state as the accepting state. For every state without an exiting transition, add a transition from that state to the accepting state (meaning that the accepting state loops back to itself).

4. Label each transition leaving state A with the conjunction of propositions in A.

5. Remove transitions labeled with an ε by treating any two states connected by an ε transition as one state.

EXAMPLE 4.6 Let us continue with the last example to build the NFA for the RE $(a* \cdot b)*$. We already computed its cover as

$$a \wedge N(a* \cdot b \cdot (a* \cdot b)*) \ \vee \ b \wedge N((a* \cdot b)*) \ \vee \ \varepsilon,$$

which adds two next-time obligations. Because the obligation $N((a* \cdot b)*)$ asks for the satisfaction of the original RE whose cover has just been computed, we only need to compute the cover for $a * \cdot b \cdot (a * \cdot b)*$, as required by the other obligation. We then have

$$
\begin{aligned}
& a* \cdot b \cdot (a* \cdot b) * \\
= \ & (a \cdot a* \mid \varepsilon) \cdot b \cdot (a* \cdot b)*) \\
= \ & (a \cdot a* \cdot b \mid b) \cdot (a* \cdot b) * \\
= \ & (a \cdot a* \cdot b \cdot (a* \cdot b)* \mid b \cdot (a* \cdot b) *
\end{aligned}
$$

which gives the cover

$$a \wedge N(a* \cdot b \cdot (a* \cdot b)*) \ \vee \ b \wedge N((a* \cdot b)*).$$

Note that the cover of both of the next-time obligations above are already computed. Therefore, the cover computation is complete. Table 4.6 gives the set

Table 4.6.   States and transitions (Example 4.6).

Label	State	Next states
s1	$a, N(a * \cdot b \cdot (a * \cdot b)*)$	s1, s2
s2	$b, N((a * \cdot b)*)$	s1, s2, s3
s3	ε	

of states and their transitions. The initial states are $s1$, $s2$, and $s3$, which form
the cover of the original formula, and $s3$ goes to the accepting state with an
ε transition, meaning that the two states are collapsed. The resulting NFA is
shown in Figure 4.6.   □

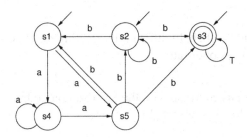

Figure 4.6.   NFA for $(a * \cdot b)*$ (Example 4.6).

To conclude this section, we give another example demonstrating the construction of an NFA for an FL$^-$ formula that consists of both LTL and RE.

EXAMPLE 4.7   Let us build the NFA for the formula $!p \rightarrow G((a \mid b)*)$. First,
we compute the cover of this formula. Note that the equality $p \wedge G(p) = G(p)$ is
used to simplify the computation.

$$
\begin{aligned}
& !p \rightarrow G((a \mid b)*) \\
= \ & p \vee (a \mid b)* \wedge N(G((a \mid b)*)) \\
= \ & p \vee ((a \mid b) \cdot (a \mid b)* \mid \varepsilon) \wedge N(G((a \mid b)*)) \\
= \ & p \vee (a \cdot (a \mid b)* \mid b \cdot (a \mid b)* \mid \varepsilon) \wedge N(G((a \mid b)*)) \\
= \ & p \vee a \wedge N(G((a \mid b)*)) \vee b \wedge N(G((a \mid b)*)) \vee \varepsilon \wedge N(G((a \mid b)*))
\end{aligned}
$$

The cover of $G((a \mid b)*)$ is already computed as a subset of the above cover, that is

$$G((a \mid b)*)$$
$$= a \wedge N(G((a \mid b)*)) \vee b \wedge N(G((a \mid b)*)) \vee \varepsilon \wedge N(G((a \mid b)*)).$$

No new obligation is produced, and so the cover computation is complete. The states and transitions are given below. The initial states are $s1$, $s2$, $s3$, and

Table 4.7. States and transitions (Example 4.7).

Label	State	Next states
s1	p	
s2	a, N(G(a|b)*)	s2, s3, s4
s3	b, N(G(a|b)*)	s2, s3, s4
s4	ε, N(G(a|b)*)	s2, s3, s4

$s4$, which constitute the cover of the original formula; we add an accepting state which has a self-loop and takes a transition from $s1$. This time, we show both the NFA with the ε transitions, in Figure 4.7, and the NFA without the ε transitions, in Figure 4.8. □

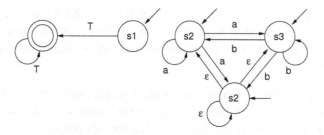

Figure 4.7. NFA for $!p \to G((a \mid b)*)$ with ε-transitions (Example 4.7).

4.5.3 Determinization of NFA

A word may cause multiple runs in an NFA because one letter may enable many transitions from any state. Determinization of an NFA, however, incurs an exponential state space increase in the worst case. HDL simulators are inherently deterministic. To use NFAs directly as monitors or generators, a simulator has

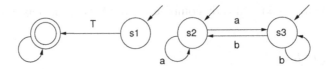

Figure 4.8.   NFA for $!p \rightarrow G((a \mid b)*)$ without ε-transitions (Example 4.7).

to determinize the simultaneous runs one way or another. Therefore, converting the NFA to a DFA seems to be wise choice, given that it avoids the run-time overhead during simulation.

The basic idea of subset construction is this: if a state goes to multiple states on one letter in an NFA, then define its new next state as the set of all these states. Eventually, in the DFA, every state is a subset of the states in the NFA. The following is the subset construction algorithm.

1. Add the set of initial states of the NFA as the initial state of the DFA;

2. For every newly added DFA state s:

>   For each letter, l, in the alphabet, compute the next state, t:

>>   For each NFA state in s, put all the NFA states that result from the application of l into t.

>>   If t is not already added, add t to the NFA; add the transition s to t and label the transition with l;

3. The accepting states of the DFA are those which contain any of the accepting states of the NFA.

The DFA state that is an empty set (containing no NFA states) can be understood as a rejecting state because its only exiting transition is a self-loop. Once reaching this state, the run gets stuck and has no chance of reaching the accepting state.

EXAMPLE 4.8   Table 4.8 gives the result of subset construction for the NFA shown in Figure 4.8. Each line in the table corresponds to a transition in the DFA. *Input* stands for the combination of letters (or propositions), A for the accepting state and T for all possible input combinations. The corresponding DFA is shown in Figure 4.9. Note that $!p(a|b)$ stands for $!p \wedge (a \vee b)$.

$\square$

4.6   Summary

We have just discussed issues in assertion languages and have given an in-depth analysis of the PSL language. We started with a brief survey of the main assertion languages in use, then introduced regular expressions and temporal logics, which form the foundation of PSL. Afterwards, we analyzed the strong

Table 4.8. Subset construction for the NFA of $!p \rightarrow G((a \mid b)*)$ (Example 4.8).

Label	DFA state	Input	Next state
s1	$\{p, a, b\}$	$p \wedge (a \vee b)$	$\{A, a, b\}$
	$\{p, a, b\}$	$!p \wedge (a \vee b)$	$\{a, b\}$
	$\{p, a, b\}$	$!p \wedge !a \wedge !b$	$\{\}$
	$\{p, a, b\}$	$p \wedge !a \wedge !b$	$\{A\}$
s2	$\{A, a, b\}$	$a \vee b$	$\{A, a, b\}$
	$\{A, a, b\}$	$!a \wedge !b$	$\{A\}$
s3	$\{\}$	T	$\{\}$
s4	$\{a, b\}$	$a \vee b$	$\{a, b\}$
	$\{a, b\}$	$!a \wedge !b$	$\{\}$
s5	$\{A\}$	T	$\{A\}$

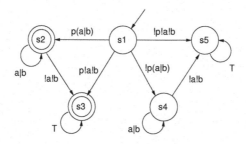

Figure 4.9. DFA for $!p \rightarrow G((a \mid b)*)$ without ε-transitions (Example 4.8).

and weak semantics of PSL and their implications on the truthness of properties on finite, truncated paths. We also covered the safety/liveness classification of temporal properties. Following the discussions of these key concepts, we covered the syntax of the PSL language. In the remainder of the chapter, we explained the assertion/constraint and the corresponding monitor/generator dualities. These dualities were highlighted in the discussion of a procedure that converts a property in a subset of PSL to a finite state machine and a set of propositions. The same finite state machine and propositions can be used either as an assertion or as a constraint, and in almost identical ways. This conversion also enables us to focus only on the propositional constraints in the discussions of test generation in later chapters.

Chapter 5

PRELIMINARIES

In this chapter, we define various notations and terminologies. First, we give a quick review of Boolean algebra and functions, and then show how digital hardware designs are modeled using them. Next, we describe how reachability analysis, the basis for many formal verification methods, is performed in the Boolean representation of designs. Afterwards, we give a detailed exposition of the symbolic data structure Binary Decision Diagrams (BDDs), which are used extensively in the constraint solving methods to be discussed later in this book. Finally, we cover the main concepts of the Boolean Satisfiability (SAT) solving, which provides an alternative to BDD-based constraint solving methods.

5.1 Boolean Algebra and Notations

A *Boolean Algebra* is defined by the set $\mathbb{B} = \{0, 1\}$ and two Boolean operations, denoted by $+$, the *disjunction*, and $\cdot$, the *conjunction*, which satisfy the *commutative* and *distributive* laws, and whose identity elements are 0 and 1, respectively. In addition, any element $a \in \mathbb{B}$ has a complement, denoted by $\bar{a}$, such that $a + \bar{a} = 1$ and $a \cdot \bar{a} = 0$.

A Boolean variable is one whose range is the set $\mathbb{B}$. A *literal* is an instance of a Boolean variable or of its complement. The multi-dimensional space spanned by n Boolean variables is the n-nary Cartesian product $\mathbb{B} \times \ldots \times \mathbb{B}$, denoted by $\mathbb{B}^n$. A point in this Boolean space is represented by a *vector* of dimension n. When Boolean variables are associated with the dimensions, a point can be identified by the values of the corresponding variables, i.e., by a *minterm*, the product of n literals. Often, products of literals are called *cubes* since they can be graphically represented as hypercubes in the Boolean space.

A Boolean formula is a composition of Boolean variables, the constants 0 and 1, and Boolean operations, that obey the following rules:

- 0, 1, and Boolean variables are Boolean formulas

- if f is a Boolean formula then so is $\overline{f}$

- if f and g are Boolean formulas then so are $f + g$ and $f \cdot g$

Many other Boolean operations can be used to abbreviate formulas generated above; for example, $f \oplus g$, where $\oplus$ denotes the *exclusive-OR*, denoted by $\oplus$, is a shorthand for $f \cdot \overline{g} + \overline{f} \cdot g$.

Since all Boolean formulas evaluate to either 0 or 1, they define mappings from $\mathbb{B}^n$ to $\mathbb{B}$, where n is the number of variables in the formula. Such mappings are called the Boolean functions. We will use the terms Boolean formula, and Boolean function, or simply *function*, interchangeably, as deemed appropriate by the context.

Given a function $f : \mathbb{B}^n \mapsto \mathbb{B}$, the set of minterms $\{\alpha \in \mathbb{B}^n \mid f(\alpha) = 1\}$ is called the *onset* of f, denoted by f^{on}, and the set $\{\alpha \in \mathbb{B}^n \mid f(\alpha) = 0\}$ is called the *offset* of f, denoted by f^{off}. Conventionally, a function f, when treated as a set, represents f^{on}. For this reason, f is called the *characteristic function* of the set f^{on}. Conversely, a set S, when treated as a function, represents the characteristic function of S.

The *cofactor* of $f(x_1, \ldots, x_i, \ldots, x_n)$ with respect to x_i, denoted by $f\mid_{x_i=1}$ or simply f_{x_i}, is computed as $f(x_1, \ldots, 1, \ldots, x_n)$; similarly, the *cofactor* of $f(x_1, \ldots, x_i, \ldots, x_n)$ with respect to $\overline{x_i}$, denoted by $f\mid_{x_i=0}$, or simply $f_{\overline{x_i}}$, is computed as $f(x_1, \ldots, 0, \ldots, x_n)$. The notation f_c, where c is a cube, represents the successive application of cofactoring of f with respect to the literals in c.

The *existential quantification* of f with respect to a variable x is $f_x + f_{\overline{x}}$, denoted by $\exists_x f$; the *universal quantification* of f with respect to x is $f_x \cdot f_{\overline{x}}$, denoted by $\forall_x f$; the *Boolean differential* of f with respect to x is $f_x \cdot \overline{f_{\overline{x}}} + \overline{f_x} \cdot f_{\overline{x}}$, abbreviated as $f_x \oplus f_{\overline{x}}$, or simply $\partial f / \partial x$.

5.2   Graphs

A graph $G(V, E)$ is defined by the pair (V, E), where V is a set of vertices, or nodes, and E is a set of edges. Each edge in E is a pair of vertices, (s, t), such that $s, t \in V$. A *subgraph* of a graph $G(V, E)$ is a graph $G(V', E')$ such that V' is a subset of V and E' is a subset of E. An *undirected graph* is a graph whose edges are unordered pairs of vertices. A *directed graph* is a graph whose edges are ordered pairs of vertices. A *path* is a sequence of vertices $s_1, s_2, \ldots, s_n)$ such that (s_i, s_{i+1}) for $1 \le i \le n - 1$ are distinct edges. We say this path is from s_1 to s_n and has a length of $n - 1$. A *cycle* is a closed path, i.e., a path going from a node back to itself. A graph is *acyclic* if it contains no cycles. A graph is *connected* if for any two nodes s and t, there is a path from s to t or from t to s. A *cut* is either a set of edges, or a set of nodes, whose removal from a connect graph disconnects the graph.

A connected acyclic graph is called a *tree*. For any two nodes s and t in a tree, there is a unique path from s to t and a unique path from t to s, and the two paths are the reverse of each other. A tree has one node designated as the *root*. Given a root r, a set of nodes in a tree are called the leaves if for each node s in the set, the unique path from r to s cannot be extended.

A directed acyclic graph is called a *dag*. We say a directed edge (s, t) (from s to t) is outgoing with respect to s and incoming with respect to t (or incident to t). The roots of a dag are the nodes that have no incoming edges. The leaves of a dag are the nodes that have no outgoing edges. The vertex pair in an edge of a tree can be ordered, say as in the pair (s, t) (from s to t), such that the unique path from the root to t goes through s. In an "ordered" tree or a dag, we say s is a *parent* of t and t is a *child* of s if (s, t) is an (ordered) edge, and we say s is an *ancestor* of t and t is a *descendant* of s if there is a path from s to t.

In a graph, we say the vertex t is reachable from the vertex s if there is a path from s to t. We define all vertices to be reachable from themselves. Given a set of vertices I, the set of vertices R that are reachable from I can be computed by either a Breadth-First-Search (BFS) or a Depth-First-Search (DFS) in the graph.

A BFS maintains a frontier set (of vertices), F, initialized to I. Each BFS step adds the vertices in F to R and replaces F with a new frontier F' such that F' and R are disjoint and every vertex in F' is reachable from a vertex in F with a path of length 1. The search terminates if the new frontier is empty.

A DFS keeps a stack of to-be-explored vertices. Initially, the stack consists of vertices from I. At each DFS step, the top vertex in the stack is examined. If the vertex reaches another vertex not in R through a path of length 1, then the reached vertex is pushed to the stack. Otherwise, the top vertex is popped and added to R. The search terminates when the stack is empty.

5.3   Hardware Modeling

As discussed in Chapter 1, hardware modeling can be done at different levels. The most dominant are the RT and gate levels. Most functional verification tools take models from these two levels as inputs. Internal to those tools, however, RTL models are synthesized into gate netlists that are more amenable to logical reasoning. For tools that handle only binary values, the unknown X and high impedance Z values need to be coerced into the 2-valued system. For example, some systems treat X as a free variable and Z as 0 or 1 arbitrarily or according to the "pullup" and "pulldown" options in Verilog. Also, for buses with multiple drivers, priority encoding circuitry may be used as the single driver. The validity of these conversions are based on certain assumptions, such as the Z values are never sampled and there are no bus contentions. In verification tools, synthesis and the conversion to Boolean logic are usually referred to as the modeling problem. Assumptions made in modeling need to be independently validated.

The resulting models are usually *bit-blasted*, that is, bits from buses and vectors are assigned and used separately as if they were scalar signals. Recently though, in sync with the trend of high level verification, *word-level modeling* that deals with buses and vectors in their entirety has caught the attention of the EDA industry. Tools based on word-level models are expected to be more scalable and more natural to the end users. However, they must overcome the challenge of shifting away from the more familiar domain of 0s and 1s. For modeling, word-level models are actually closer to RTL and easier to construct than their bit-level counterparts. Figures 5.1 through 5.4 give an example of bit- and word-level models.

```
always @ (posedge clk or posedge reset)
begin
    if (reset)
        full <= 1'b0;
    else if (write_ptr == read_ptr)
        full <= 1'b1;
end
```

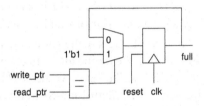

Figure 5.1.   RTL description.

Figure 5.2.   Word-level model.

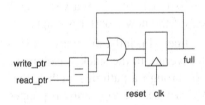

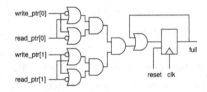

Figure 5.3.   Word-level model with optimization.

Figure 5.4.   Bit-level model.

In the remainder of this section, we will focus on the construction of the Finite State Machine (FSM) model from a bit-level netlist. An FSM model defines the behavior of a hardware design, in contrast to the structure of the design given by a netlist. FSM models are required by both static verification and constrained random simulation.

A netlist consists of an interconnected set of primary inputs, outputs, gates, and latches. Each gate performs a Boolean operation on the signal values of its inputs, and produces the result at its output. A special primary input, called a *clock*, governs the latches in a manner such that a latch either maintains the value of its output, or updates the value to that of its input if a designated *clock*

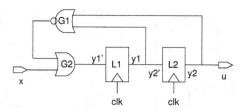

Figure 5.5. Example: A netlist.

event occurs, e.g., the clock changes from 0 to 1. Consequently, a latch is often referred to as a *storage* or *state-holding* element, and is identified by a state variable representing its output.

Let $X = \{x_1, \ldots, x_n\}$ be the set of variables representing the primary inputs, and $Y = \{y_1, \ldots, y_m\}$ the set of state variables, where y_i represents the i-th latch. A state is a valuation of Y, given by the minterm

$$\bigwedge_{i=1}^{m} (y_i == c_i)$$

where c_i is the valuation of y_i, $y_i == 1$ and $y_i == 0$ correspond to the literals y_i and $\overline{y_i}$, respectively.

The valuation of y_i is determined by the *transition function* $F_i(X, Y)$, which is the composition of gate operations transitively affecting the input of the i-th latch. Define $Y' = \{y_1', \ldots, y_m'\}$ to be the set of *next state* variables, where y_i' represents the input of the i-th latch. The *transition relation* of y_i' is defined as

$$T_i(y_i', X, Y) = (y_i' == F_i(X, Y)).$$

For designs where all latches are synchronized by the same clock, meaning that they transition at the same time, a global transition relation can be defined as

$$T(Y', X, Y) = \bigwedge_{i=1}^{m} T_i(y_i', X, Y).$$

Separately, the outputs of the design are determined by a vectorial function $U(X, Y)$, the composition of gate operations that drive the outputs.

The transition relation and output function can be visualized in an FSM that provides a behavioral description of the design. An FSM is an edge-labeled directed graph $G = \{V, E, L\}$, where the vertices V correspond to the *states*, the edges E correspond to the state transitions, and the labels L represent the *input-output* pairs. More precisely, we have

$$V = \mathbb{B}^{|Y|} = \mathbb{B}^m$$

x	y_1	y_2	y_1'	y_2'
0	0	0	1	0
0	0	1	0	0
0	1	-	0	1
1	0	-	1	0
1	1	-	1	1

"-" means "0 or 1" .

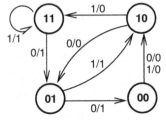

"0/1" means the input is 1, the output is 0.

Figure 5.6.   Example: Transition relation and finite state machine.

$$E = \{(s, t) \mid \exists x (t, x, s) \in T$$

$$L = \{(s, t, x, u) \mid (t, x, s) \in T, u = U(x, s)$$

That is, if an edge $s \xrightarrow{(x,u)} t$ exists in the graph, then applying the input x at the state s results in an output of u, and a transition to the state t.

EXAMPLE 5.1   Consider a design whose netlist is given in Figure 5.5. The design has two latches, $L1$ and $L2$, an OR gate, $G1$, and a NOR gate, $G2$. The transition functions are $y_1' = x + \overline{y_1 + y_2}$ and $y_2' = y_1$, and the output function is $u = y_2$. The corresponding transition relation and FSM are shown in Figure 5.6.   □

5.4   Reachability Analysis

A common functional verification problem, known as *invariant checking*, is to decide whether some designated "bad" states are *reachable* from a set of initial states of the design. This kind of problem can often be solved by the *reachability analysis*, a method for state space exploration that is the basis of model checking.

Given the FSM of a hardware design, a state t is said to be reachable from s if there is a sequence of transitions that starts from s and ends at t. The set of states reachable from s can be obtained either by a DFS or by a BFS in the FSM. In comparison, a DFS explores one successor state per step, thus requiring less memory, but a longer run time in general, than a BFS, which explores all the successor states at each step. A natural compromise is to always start with a BFS if the computation resources allow, otherwise exploring as many states as possible, per step. We will therefore give more details only on the BFS-based analysis.

In the previous section, we showed that one can build the FSM of a design from its transition relation. The construction is actually a one step BFS-based state traversal, starting from all the states. Similarly, reachability analysis on an FSM can be implicitly done using the transition relation, as in the following.

```
Reachable(R⁰) {
    R = R⁰;
    ΔR = R⁰;
    do
        R' = Img(ΔR);
        ΔR = R' ∩ ¬R;
        R = R ∪ ΔR;
    while (ΔR ≠ ∅);
    return R;
}
```

Figure 5.7. Reachability Analysis.

Let $T(Y', X, Y)$ be the transition relation, where X, Y and Y' are the input, state and next state variables, respectively. Let $C(X, Y)$ be the constraint. The set of states reachable in one step from the states in R is the *image* of R under T subjected to C, or formally

$$Img(R(Y)) = \exists_{XY}[R(Y) \cdot T(Y', X, Y) \cdot C(X, Y)]. \tag{5.1}$$

The series $R^0, R^1, \ldots$ where $R^{i+1} = Img(R^i) \vee R^i$ increases monotonically. Since the state space is finite, the series has a *least fixed point (lfp)* R^K where K is finite and for all $i \geq K$ $R^{i+1} = R^i$. Using the *lfp* operator (μ) from Mu-Calculus [Par70, EC81], the computation of R^K can be characterized as

$$\mu Z. [Img(R^i) \vee Z]. \tag{5.2}$$

Figure 5.7 gives the corresponding algorithm with the optimization that *Img* is performed only on states that are "new" in R^i.

For a simple illustration of the algorithm, in the FSM shown in Figure 5.6(b), the reachable states of the state 11 is computed in the series $\{\{11\}, \{11, 01\}, \{11, 01, 00\}, \{11, 01, 00, 10\}\}$.

Invariant checking can also be approached by the *backward reachability analysis*, which computes the set of states that reaches, instead of being reachable from, the bad states. The recursion is based on the following *pre-image* operator

$$PreImg(R(Y')) = \exists_{XY'}[R(Y') \cdot T(Y', X, Y) \cdot C(X, Y)], \tag{5.3}$$

which also leads to a *lfp* computation

$$\mu Z. [PreImg(R^i) \vee Z]. \tag{5.4}$$

The algorithm implementing the above would be identical to that of the (forward) reachability analysis shown in Figure 5.7, except that Img would be replaced by $PreImg$.

5.5   Reduced Ordered Binary Decision Diagrams

Th binary decision diagram BDD was first introduced by Lee [Lee59] in 1959 and Akers [Ake78] in 1978 as a compact representation for Boolean functions. In the mid-1980s, Bryant [Bry86] proposed Reduced Ordered Binary Decision Diagrams (ROBDDs) by imposing restrictions on BDDs such that the resulting representation is canonical. He also presented efficient ROBDD algorithms for Boolean operations. Since Bryant's work, related research has flourished, mainly in the field of formal verification, but also in logic synthesis. In this chapter, we give an overview of BDDs and ROBDDs.

5.5.1   BDD Representation of Boolean Functions

A BDD is a rooted, dag wherein each internal node has two sub-BDDs and the terminal nodes are *ONE* and *ZERO*, representing the Boolean values 1 and 0, respectively. A BDD γ identifies a Boolean function f of variables X, for a given mapping l from its nodes to X, as in the following.

1. Let r be the root of γ

2. If r is *ONE*, then $f = 1$

3. If r is *ZERO*, then $f = 0$

4. If r is an internal node, let $v = l(r)$ be the associated variable, g and h be the functions identified by r's *right* and *left* sub-BDDs, respectively, then $f = v \cdot g + \overline{v} \cdot h$

Because of the composition in Item 4, above, we sometimes use the adjectives *then* and *else*, in place of *right* and *left*, in distinguishing between the two subgraphs of a BDD node.

An ROBDD is an ordered and reduced BDD, as defined in the following:

1. An ordered BDD is a BDD whose variables follow a total ordering $\prec$ such that $l(s) \prec l(t)$, if t is a descendant of s.

2. An ROBDD is the maximally reduced version of an ordered BDD obtained by repeatedly applying to the ROBDD the following rules until none are applicable.

    a. If two subgraphs are identical, remove one of them, and let the dangling edge point to the remaining subgraph

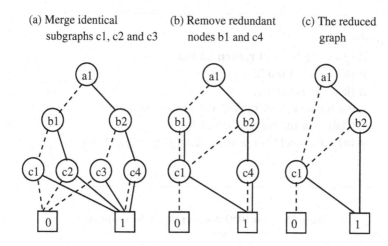

(a) Merge identical
subgraphs c1, c2 and c3

(b) Remove redundant
nodes b1 and c4

(c) The reduced
graph

Figure 5.8. Reducing BDD for function $a \cdot b + c$.

b. If a node points to the same subgraph, remove the node, and let the dangling edge point to the subgraph

Figure 5.8 illustrates the derivation of an ROBDD from the "complete" BDD of function $a \cdot b + c$. The index $b2$ means that the node is the second node labeled with variable b.

ROBDD is the symbolic data structure used extensively this book. For brevity, we will refer to ROBDDs simply as BDDs hereafter.

5.5.2 BDD Manipulations

All Boolean operations on BDDs can be implemented using one procedure called *Apply* [Bry86], based on the Shannon expansion

$$g \circ h = v \cdot (g_v \circ h_v) + \overline{v} \cdot (g_{\overline{v}} \circ h_{\overline{v}}) \tag{5.5}$$

where $\circ$ is a Boolean operation and v a variable in the support of g or h. Let $Compose(v, l, r)$ be the BDD that is a composition of variable v and two BDDs l and r, such that the root is labeled with v and the left and right sub-BDDs are l and r, respectively. Let *Reduce* be the reduction procedure given in the previous section. *Apply* recursively constructs a BDD from two operant BDDs and a binary operator, based on the following recursion, a direct translation of Equation (5.5).

$$Apply(g \circ h) = Reduce(Compose(v, Apply(g_v \circ h_v), Apply(g_{\overline{v}} \circ h_{\overline{v}})))$$

As an optimization, *Apply* designates v as the higher-ranked variable of the variables labeling the roots of g and h. The terminal cases of the recursion

```
Apply(g · h) {
    if (g == 0 || h == 0) return ZERO;
    if (g == 1) return h;
    if (h == 1) return g;
    Let v be the higher-ranked variable of variables
        labeling the roots of g and h;
    return Reduce(Compose(v, Apply(g_v · h_v), Apply(g_v̄ · h_v̄)));
}
```

Figure 5.9.   The BDD Apply operation for conjunction.

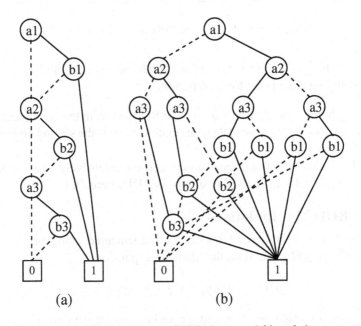

(a)   (b)

Figure 5.10.   Dependency of BDD size on variable ordering.

depend upon the selection of ∘. For instance, Figure 5.9 gives the *Apply* function for the BDD conjunction operation. Note that the complement of f is computed as $Apply(f \oplus 1)$.

5.5.3   The BDD Size Consideration

For a given function, the size of its BDD representation, that is, the number of nodes in the BDD, varies with the variable ordering. To demonstrate this dependency, we use an example taken from [Bry86]. Assume we are to construct a BDD for the function

$$f = a_1 \cdot b_1 + a_2 \cdot b_2 + a_3 \cdot b_3$$

With the ordering $a_1 \prec b_1 \prec a_2 \prec b_2 \prec a_3 \prec b_3$, i.e., variables appear in the order $a_1 b_1 a_2 b_2 a_3 b_3$ on any path from the root to a terminal node. The BDD representation of f is shown in Figure 5.10(a). It has 8 nodes. In contrast, with $a_1 \prec a_2 \prec a_3 \prec b_1 \prec b_2 \prec b_3$, the BDD representation as shown in Figure 5.10(b) has 16 nodes. The first ordering is sometimes referred to as the *interleaved* ordering of the a and b variables. More generally, for the function $a_1 \cdot b_1 + \cdots + a_n \cdot b_n$, the interleaved ordering yields a BDD with $2(n + 1)$ nodes, whereas the second ordering yields a BDD with 2^{n+1} nodes. Evidently, a poor choice of variable orderings can have very undesirable effects.

Thanks to pioneering works such as the static ([FS90]) and dynamic ([Rud93]) variable orderings, a variety of techniques and heuristics [BRKM91, ATB94, FOH93, FYBSV93, PS94, BLW95, DBG95, FAA01] have since been proposed and shown to be effective in finding good variable orderings for BDDs. In practice, good variable orderings exist for a large class of Boolean functions, and their BDDs are much more compact than traditional representations such as truth tables, and conjunctive and disjunctive normal forms. For this reason, BDDs are widely used in design automation tools where the compact and canonical representation of functions is in order.

Before we conclude this section, we need to mention another common problem with BDD sizing. During a BDD operation, the intermediate BDDs can have a much larger size than the final BDD. This is especially noticeable when there are many BDD operants, for example, in the conjunction of a large group of BDDs. Techniques such as *conjunctive partitioning* [RAP$^+$95] and *early quantification* [HKB96] have been developed to handle this problem.

5.6 Boolean Satisfiability

Boolean Satisfiability (SAT) is the problem of finding the answer to the following question: given a Boolean expression, is there an assignment of true and false values to the variables that will make the entire expression true?

SAT belongs to the class of NP-complete problems whose algorithmic solution is believed to have an exponential worst case complexity [Coo71]. In practice, however, SAT solvers are found to be quite efficient for many applications in Artificial Intelligence, Operational Research, and EDA (timing analysis, FPGA routing, and logic synthesis, to name a few). Recent enhancements to the SAT solving capacity [MSS96, MMZ$^+$01, GN02] have been the main driving force behind the latest development in formal verification (e.g., [BGCZ99, McM03, MA03, GGYA03]).

5.6.1 SAT Solving

A naïve approach to SAT solving is to try all possible variable assignments and see if any of them satisfy the formula. This immediately leads to an exponential time complexity. Unfortunately, this exponentiality characterizes SAT as being in the category of NP-Complete problems: a worst-case exponential time complexity is believed to unavoidable. In fact, as proved by Stephen Cook in 1971 [Coo71], SAT was the first known NP-complete problem. Despite this theoretical limit, SAT algorithms with good heuristics and randomized SAT solvers are effective in practice. In fact, the construction of BDDs, discussed earlier, is a SAT solving process. Unsatisfiable formulas correspond to the zero BDD, that is, a BDD with just the ZERO node. In a non-zero BDD, each path from the root node to the ONE node represents a satisfying assignment to the corresponding formula. In the sequel, we shall restrict ourselves to an introduction of logic deduction and search-based algorithmic approaches to SAT.

The earliest SAT algorithm was developed by Davis and Putnam [DP60]. Their algorithm iteratively selects variables to resolve until no more variables are left (the problem is satisfiable) or a conflict is encountered (the problem is unsatisfiable). This is equivalent to the existential quantification of variables, and is exponential in memory usage.

Later, Davis, Logemann, and Loveland (DLL) [DLL62] proposed a DFS-based algorithm with backtracks. For historical reasons, it is also known as the Davis-Putnam-Logemann-Loveland (DPLL) algorithm. Most modern, high-performance SAT solvers are variations of DPLL. A typical SAT solver relies on a few key procedures:

- *Decides*: makes a *decision* by picking an unsigned variable and assigning it a value (0 or 1). It returns false iff there are no more unassigned variables.

- *Deduces*: propagates in the clauses the decision assignment and assignments implied by propagation. It returns false iff these assignments conflict.

- *Analyzes*: determines, when a conflict occurs, which decision to backtrack. Optionally it learns new information from the conflict.

- *Backtrack*: undoes assignments up to a specified decision level. A new assignment is made to branch the search.

Figure 5.11 shows how these procedures work together.

As simple as it looks, this algorithm has been the subject of continuing research for the past forty years. In fact, each of the main procedures offers a unique opportunity for improvement. The order in which a variable is picked, and the value assigned in the decision, can affect the actual search space. The same is true with conflict analysis and backtracking. Backtracks can be made non-chronologically (that is, not just backtrack to the last decision) with an

```
SAT() {
    dlevel = 0;
    while (true)
        if (deduce() == false)
            dlevel = analyze();
            if (dlevel == 0) return UNSAT;
            else backtrack(dlevel);
        else if (decide() == false) return SAT;
        else dlevel = dlevel + 1;
}
```

Figure 5.11.   A typical SAT algorithm.

intelligent analysis of the conflict. Also, the efficient implementation of the heavily used deduction/propagation procedure can greatly improve the overall performance of the algorithm. We will discuss these issues in greater detail after giving a few useful definitions.

5.6.2   Definitions

SAT problems become easier to reason about when formulated in simple and regular representations. The most popular choice of representation has been the Conjunctive Normal Formal (CNF). CNF is a conjunction of clauses. Each clause is a disjunction of literals. A literal is a variable (thus a positive literal) or the negation of a variable (thus a negative literal). Converting a (gate-level) netlist to a CNF formula is straightforward: each gate translates to a set of at most three clauses. For example, an AND gate $a = b \& c$ defines the relation $a \rightarrow b \& c$ and $b \& c \rightarrow a$, which gives three clauses $(\bar{a} + b)$, $(\bar{a} + c)$, and $(a + \bar{b} + \bar{c})$. Four clauses are obtained for an XNOR gate using same the approach. Note that a straightforward composition of clauses for the three 2-input gates of XNOR would have given nine clauses. This shows the effect of collapsing when clauses are derived directly from the function instead of from the constituent gates. The clauses for the AND, OR, NOT, and XNOR gates are shown in Table 5.1. Also note that the clauses for the NOT gate can be eliminated by propagating the equality $a = \bar{b}$ so that the variable b (or a) is removed from all other clauses.

Given an SAT problem as a set of clauses, SAT solving is concerned with finding a assignment to the variables such that all clauses are satisfied. A clause is satisfied iff it has at least one *true literal*. A true literal is a literal that, given the assignment to its variable, evaluates to 1. That is, if variable $x = 1$ (resp.

Table 5.1.   Clauses of the AND, OR, NOT, and XNOR gates.

Gate	Definition	Clauses
AND	$a = b \cdot c$	$(\bar{a} + b)(\bar{a} + c)(a + \bar{b} + \bar{c})$
OR	$a = b + c$	$(a + \bar{b})(a + \bar{c})(\bar{a} + b + c)$
NOT	$a = \bar{b}$	$(a + \bar{b})(\bar{a} + b)$
XNOR	$a = (b + \bar{c})(\bar{b} + c)$	$(a + b + c)(a + \bar{b} + \bar{c})(\bar{a} + \bar{b} + c)(\bar{a} + b + \bar{c})$

$x = 0$), then x (resp. $\bar{x}$) is a true literal. A *false literal* is the negation of a true literal. A literal is *free* if its variable is not assigned. A literal is pure with respect to a set of clauses if it appears in at least one of these clauses and its negation does not. A pure literal can be immediately assigned to true so that the clauses containing it are satisfied and removed from consideration.

An unsatisfied clause can be found in two interesting states. A clause is called a *conflicting* clause if all of its literals are false. A clause is called a *unit* clause if exactly one of its literals is free and all others are false. To satisfy a unit clause, the only free literal must be made true. This imposes a corresponding assignment to the variable of the free literal. We call the assignment an *implication*, and the unit clause its *antecedent*.

Contrary to implication, a *decision* is an elective variable assignment that is not forced by any condition. SAT solving is, in a nutshell, an iterative decision-making and testing process. The most commonly used testing procedure is *Boolean Constraint Propagation* (BCP). BCP propagates a variable assignment in clauses by making free literals true or false, thereby changing the state of the clauses. Two types of resulting clauses are of interest: conflicting clauses that terminate the BCP and cause a backtrack, and unit clauses that add implied assignments and prolong the BCP.

All variable assignments are associated with a *decision level* (*dl*). The *dl* of a decision assignment reflects the depth at which the decision is made in the DFS tree. The assignment at the root of the tree has a *dl* of 1. Each decision increments the *dl* by 1. Implied assignments from a BCP are associated with the same *dl* as the decision that initiated the BCP. A backtrack cancels all decisions and implied assignments between the *dl* and a given *backtrack level*, disassociates these assignments with their *dl*s, and sets the decision level (lower) to the backtrack level.

Implications made in a BCP can be captured in an *implication graph*. An implication graph is a DAG defined as follows.

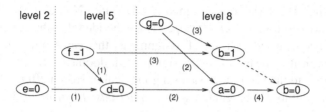

Figure 5.12. An example of BCP.

- Each leaf node in the graph is either a decision assignment or an implied assignment from lower *dls* (than the BCP considered), and each internal node is an implied assignment in the BCP.

- Each edge, from node A to node B, is labeled by a clause c such that c is the antecedent of B and A is a false literal of c. Additionally, if there are other edges incident to B, they must also be labeled as c.

If nodes A and B are conflicting assignments, where A is assigned before B, we add a dummy, dotted edge from A to B to signify the conflict. We call B the *conflicting node* and the antecedent clause of B the *conflicting clause*.

EXAMPLE 5.2 For an example of BCPs, decision levels and conflicts, consider the following clauses.

$$
\begin{array}{llll}
(1) & (\bar{d}+e+\bar{f}) & (4) & (a+\bar{b}) \\
(2) & (\bar{a}+d+g) & (5) & (b+e+\bar{g}) \\
(3) & (b+\bar{f}+g) & (6) & (\bar{a}+\bar{f}+g)
\end{array}
\tag{5.6}
$$

Suppose the decisions $e = 0$, $f = 1$, and $g = 0$ have been made at the *dls* 2, 5, and 8, respectively. The assignment $d = 0$ belongs to *dl* 5 since it is the result of the BCP of $f = 1$. Let the current *dl* be 8. The current BCP leads to the implications $b = 1$, $a = 0$, and $b = 0$, at *dl* 8, which contain a conflict at b. Assume $b = 1$ is implied before $b = 0$, then the conflicting node is $b = 0$ and the conflicting clause is (4). It is also possible to imply $b = 0$ first, in which case, the conflicting node and clause are $b = 1$ and (3), respectively. Note that the clauses (5) and (6) are not in the graph, and therefore do not contribute to the conflict. □

5.6.3 Conflict-based learning and Backtracking

Backtracks in the early SAT algorithms are characterized as being *chronological*. This means that when conflicts occur, the search goes backward on the ordered list of assigned variables and resumes with the *first* variable with an untried value. The variable is assigned to that value. Non-chronological back-tracking may go further back. It depends on the idea that the current conflict may have been caused by a set of assignments that came much earlier. Any search attempt after the last of such assignments is doomed to failure. Therefore, jumping back to the right place saves a good deal of search effort.

This back-jump can be guided with *conflict-based learning* (clause record-ing). Non-chronological backtracking using *nogood* recording was originally proposed by Stallman and Sussman in [SS77] and further studied in [Gas79] and others. The idea is extended to a CNF-based SAT solver in [MSS96] and [JS97].

Marques-Silva and Sakallah in [MSS96] presented a tool called *GRASP* that uses implication graphs to learn clauses. Each learned clause has the following properties: it is a logical consequence of the original clauses, and it prohibits the same conflict assignment in future search. The first property is the basis for the soundness of clause recording — learned clauses only add redundancy to the original clauses. However, too much conflict-avoiding redundancy slows down the search.[1] Plus, other than causing the current backtracking, a learned clause is not guaranteed to be useful later. Therefore, learning is usually restricted. For example, large learned clauses, which are not very useful in confining the search, are discarded; less active clauses (with many free literals), or an arbitrary percentage of learned clauses, are periodically deleted.

From an implication graph, one or many clauses can be learned. A learned clause corresponds to a *cut* in the graph, that is, a set of assignments that are sufficient to cause the conflict. Literals in the learned clause are the negations of these assignments. There can be many cuts in the graph. Heuristically, cuts that are "good" for learning contain one or more Unique Implication Points (UIPs) [MSS96]. A UIP is a node that is on all the paths from the current decision node to the conflicting node. This definition is extended in [ZMMM01]. The current decision node itself is clearly a UIP. Once a UIP is located, a cut is made such that it contains the UIP and all of the descendants of the UIP are on the same side of the cut.

Non-chronological backtracking is realized by using a learned clause that is *asserting*. A clause is asserting, at a given *dl*, if all but one of its literals are false literals assigned at lower levels than the *dl*. That remaining one literal is thus asserted (to be true). Given an asserting learned clause, non-chronological backtracking proceeds as follows:

---

[1] To an extreme, backtracking can be avoided altogether by using a maximally redundant set of clauses, defined as follows: it is the negation of the disjunction of all satisfying assignments to the negation of the original SAT problem. However, searching on this exponentially sized set defeats the purpose of avoiding backtracking.

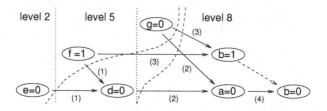

Figure 5.13. Example: BCP and learning.

1. Let the backtrack level be the maximum of the *dls* of all literals in the learned clause other than the asserted literal. This literal must be a UIP if the clause is learned with the UIP heuristic.

2. Undo all assignments between the current *dl* down to, but excluding, the backtrack level.

3. Update the current *dl* to the backtrack level.

4. Assign the asserted literal to be true at the current *dl*.

It is worth noting that the assignment to the asserted literal is an implication, not a decision. The learned clause is the antecedent of this implication. Compared to chronological backtracking, the above non-chronological backtracking not only flips the last decision automatically, but also selects a lower decision to jump back to, all driven by conflict-based learning.

EXAMPLE 5.3 We continue with the last example to illustrate conflict analysis and non-chronological backtracking. See Figure 5.13. The only UIP is the node $g = 0$. There are two possible cuts, both leading to asserting clauses. The two learned clauses, numbered 7 and 8, are given below.

$$(7) \quad (e + \overline{f} + g)$$
$$(8) \quad (d + \overline{f} + g)$$

Usually, learning one clause per conflict is sufficient. We will look at both learned clauses anyway. The two clauses cause the same backtracking: it cancels all assignments at *dls* 8, 7, and 6, changes the current *dl* to 5, i.e., the *dl* of $f = 1$, and asserts g. The ensuing BCPs under the two clauses are different, as shown in Figures 5.13 and 5.13. Nonetheless, both BCPs encounter the same conflict and learn the same clause: $e + \overline{f}$ ($f = 1$ is the new UIP). The new backtracking based on this clause decreases the current *dl* to 2, the *dl* of $e = 0$, and asserts $\overline{f}$. This BCP does not cause any conflict. As we have seen, the backtrackings made non-chronological back-jumps from the level 8 to 5 and to 2. □

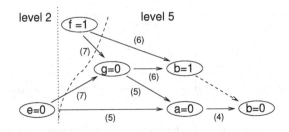

Figure 5.13 (continued).   Example: BCP and learning.

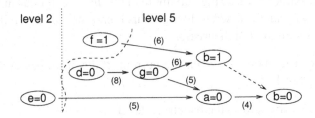

Figure 5.13 (continued).   Example: BCP and learning.

5.6.4   Decision Heuristics

Decision heuristics are responsible for the selection of variables and assignments in the *decide()* procedure shown in Figure 5.11. Similar to non-chronological backtracking, good decision heuristics contribute to the overall performance of SAT solving by reducing the search space.

The simplest decision heuristics choose variables and assignments randomly. The original DPLL algorithm selects a literal from an unsatisfied clause of minimum size. More sophisticated schemes involve maximizing certain forms of weight functions. For example, the Dynamic Largest Individual Sum (DLIS) selects a literal (to make it true) that occurs most frequently in the unsatisfied clauses. Several variations are also possible [MS99]. The comparison of decision heuristics is complicated by the interplay of decision and constraint propagation. Fewer decisions may not always result in faster solving because more work may have to be done in constraint propagation. Furthermore, characteristics of conflicts may also have a say in decision making. This is especially true for SAT solvers that adopt conflict-based learning and non-chronological backtracking. In a recent efficiently implemented SAT solver, *Chaff* [MMZ$^+$01], Moskewicz *et al.* proposed a Variable State Independent Decaying Sum (VSIDS) heuristic that emphasizes the role of conflict analysis. The VSIDS heuristic works as follows:

1. Each (positive and negative) literal of each variable has a counter, initialized to 0.

2. When a clause is learned, the counter associated with each literal in the clause is incremented.

3. The unassigned variable, and its literal with the highest counter, is chosen at each decision.

4. Ties are broken randomly, by default.

5. Periodically, all the counters are divided by a constant.

VSIDS gives preference to literals of recently learned clauses. Another SAT solver, *BerkMin* [GN02], uses a variation where scores of all literals involved in conflict analysis are incremented among other decision heuristics. A later *Chaff* implementation [MFM04] uses an enhanced version of VSIDS. This trend is consistent with popular conflict-based learning and backtracking.

5.6.5 Efficient BCP Implementation

BCP is the work horse of an SAT solver. While conflict analysis is invoked only for some decisions, every decision requires at least one, and typically many, BCP iterations.

Each BCP iteration propagates one variable assignment, taking an amount of time proportional to the number of clauses that must be checked. An immediate savings comes from the fact that we are only concerned with the clauses containing a literal of the assigned variable. Furthermore, out of these clauses, only the ones in which the concerned literal is assigned to false is of interest — they may become conflict or unit clauses. The other clauses are immediately satisfied, providing no further information.

The above strategy can still be improved. The idea is that assigning a literal to false in a clause does not always warrant a critical change, that is, causing a conflict or an implication. An elegant solution that covers all of the above situations is to watch only two literals out of each clause. This is proposed by Moskewicz *et al.*, again in *Chaff* [MMZ+01], and is referred to as the *watched-literals* approach. It should be noted that, earlier, H. Zhang in his SAT solver *SATO* [Zha97] presented a similar idea that uses head and tail literals to watch for conflict and unit clauses. The difference is that *Chaff*, unlike *SATO*, does not change the watched literals when canceling assignments.

The general idea of literal watching is that implication and conflict are possible only when the last two non-false (true or free) literals of a clause are assigned to false. Thus, we try to maintain the non-falseness of two watched literals as long as possible, until implication or conflict occurs. This 2-literal-watching scheme is implemented as follows. Initially, two literals are randomly picked from each (non-unit) clause for watching. In other words, each literal of each variable has a list of clauses in which the literal is watched. We say these clauses are the *watched clauses* of that literal. In the constraint propagation of a variable assignment, the corresponding false literal is no longer suitable to be a watched literal. Therefore, we try as much as we can to replace it in its watched clauses. Let c be a watched clause, u be the watched literal to be replaced in c, and v the other watched literal in c. Four scenarios are possible in our attempt to find the replacement for u.

1. There is a non-false literal w and $w \neq v$. Therefore, we have found a replacement for u. Delete c from u's watched clauses and add it to w''s watched clauses.

2. All non-watched literals are false, so it is impossible to replace u. Then we examine the other watched literal v for further information.

   a. If v is free, then we have a new implication, $v = 1$.

   b. If v is true, then c is satisfied and we do nothing.

   c. If v is false, then we have a conflict.

The last case is very peculiar since it only occurs when assigning v to false in c implies $u = 1$, but before $u = 1$ is propagated an earlier implication $u = 0$ is propagated to c, causing the conflict. Thus, in actual implementation, conflicts are detected a little bit earlier while queuing the assignments for propagation.

The following analysis establishes that watched literals are not affected by backtracking, and that satisfied clauses are handled automatically. First, we already know that when both u and v are false, they are false at the same *dl*. Immediately, the backtracking sets them both back to free. In addition, if only v is false, then u is true (at the end of the iterative propagation), and v is at the same level as u (case 2), or at a level no lower than u (case 3). In both cases, if backtracking occurs and cancels any assignments of u or v, then v must be one of them; thus the only false watched literal is set to free. Otherwise, since both u and v remain assigned, the clause, by being satisfied, is not watched. Hence in this last case, v being false is not relevant anyway.

5.6.6   Unsatisfiable Core

The unsatisfiability of a set of clauses is usually the result of conflicting subsets. We call those subsets the *unsatisfiable cores* (or unsat cores). With DPLL-style SAT solvers, an unsat core is simply the set of original clauses that are the antecedents of the leaf nodes in the *complete* implication graph, when the problem is determined to be unsatisfiable. By "complete" we mean that all the learned clauses (as antecedents of nodes) are replaced with the implication graphs from which they are derived. Thus, the conflict is fully explained only with the original clauses, and finding the unsat core is cast as a straightforward graph traversal for discovering the leaves. For example, the core for the conflict in Figure 5.12 contains only four out of the original six clauses (see the example in Section 5.6.3).

It is then obvious that all of the earlier implication graphs must somehow be remembered. This imposes some memory overhead, and, in extreme cases, leads to memory exhaustion. Therefore in [ZM03a], where core extraction is originally presented, Zhang and Malik proposed to store the information on a hard drive. The extracted core, however, is typically small relative to the set of original clauses. This leads to the use of unsat core extraction as automatic abstraction in SAT-based model checking [MA03, GGYA03].

Alternatively, the implication graph can be viewed as an resolution graph, and core extraction formulated as an iterative resolution that should eventually derive the empty (conflicting) clause. Based on this observation, Zhang and Malik [ZM03b] proposed an SAT solver checker: resolution is iteratively performed starting from the final conflicting clause. For each clause c being resolved, the antecedent clauses of the literals in c are iteratively resolved against c, in the reverse order in which the implications were applied. Each resolution first asserts the correctness criteria that "the two resolving clauses have exactly one pair of literals of the same variable with opposing polarity." In the end, we should arrive at an empty clause.

5.6.7 Other Optimizations

Random restarts: Random restarts are introduced mainly to deal with large variances in SAT solving performance over similar problem instances [GSK98, BMS00]. In these SAT instances, the search may progress very slowly and appear to be getting stuck in some local search spaces, even with the help clause learning. Periodical restarts help the search to leave these hard spots, thereby increasing SAT solving robustness. In general, however, solving with random restarts does not guarantee termination.

Clausification and preprocessing: Clausification is the process of converting the original SAT problem into clause form [Vel04]. Optimization is possible, for example, by collapsing internal nets to a network of multiplexers and generating fewer, albeit longer, clauses. Preprocessing further optimizes clauses by eliminating variables [SP04, EB05] through resolution, and clauses through subsumption [EB05].

Variable Probing and *Lookahead*: Both of these are for predicting variable assignments as opposed to learning from bad decisions retrospectively. Variable probing is closely tied to resolution (or consensus). For example, one can deduce that a must be 1 to satisfy $(a + b)(a + \bar{b})$ by resolving on variable b. The original Davis-and-Putnam algorithm is based purely on resolution. Two recent decision procedures can be considered intelligent ways of performing systematic resolutions: the patented *Stalmark's algorithm* [Sta92] essentially resolves on variables, and *recursive learning* [KP94] essentially resolves on clauses. However, variable probings are still more costly than conflict analysis, and it is not clear how frequently they should be invoked. An experiment by Aloul and Shakallah [AS00] suggests the use of recursive learning in the preprocessing stage.

Lookahead is a decision heuristic for variable selection. In each constraint propagation, both the true and false assignments to the current decision variable are propagated, each in a separate clause database. If there is no conflict, weights for the variables are computed. For example, in [HvM04] the weight is a function of the number of newly created clauses with two literals. The variable with the highest weight is selected as the next decision variable.

5.7   Automatic Test Pattern Generation

Automatic Test Pattern Generation (ATPG) is an algorithmic approach to test vector generation for semiconductor testing. We give an overview on the fundamental ATPG problem: test generation for combinational circuits for *Single Stuck-At* (SSA) faults. Stuck-at faults capture a major class of manufacturing defects: the voltage of a line, while expected to be able to toggle, is fixed at power or ground. The SSA fault model assumes that the circuit of concern contains a single stuck-at fault at any given time. This assumption is backed by statistics showing that most stuck-at faults can be detected by assuming that they are the only fault present.

The goal of ATPG is to generate vectors that distinguish between a good and a bad circuit. The good circuit is the original design. The bad circuit is the original design altered by imposing a single stuck-at fault at a line. A generated vector must meet the following requirements:

- *Fault sensitization*: By applying the vector to the good and bad circuits, the line in the bad circuit at which the stuck-at fault is inserted and the corresponding line in the good circuit have different values.

- *Fault propagation*: By applying the same vector, at least one output in the good circuit and the corresponding output in the bad circuit have different values.

Similar to SAT, ATPG is also a backtrack search that repeatedly applies the decision, propagation, and backtracking steps. The search space is bounded by two to the number of possible decisions. An ATPG algorithm is *complete* if it can systematically explore the entire search space. The maximum set of decision points is the set of lines in the circuit. We will see that limiting the decision making to a subset of the maximum set is a main source of performance improvement for ATPG algorithms. In the sequel, we will discuss three of the best known algorithms for combinational-circuit ATPG: the D-algorithm [Rot66, RBS67], the Path-Oriented Decision Making (PODEM) [Goe81] algorithm, and the Fanout-Oriented algorithm (FAN) [FS83].

5.7.1 The D-algorithm

The D-algorithm is the first "complete" ATPG algorithm. It unifies the treatment of the good and bad circuits by using a composite-value system. Specifically, the D-algorithm deals with five composite logic values: 0/0, 1/1, 1/0, 0/1, and x/x. The first and second value in each pair represent the value in the good and the bad circuit, respectively. The five values are shorthanded as 0, 1, D, $\overline{D}$, and x. The value x, representing an *unassigned* value, can take any of the first four values. Logical operations such as AND, OR, NOT, are carried out in this 5-valued system by applying the usual 3-valued operation separately on the good and bad values and then composing the results. For example, D + 1 = 1/0 + 1/1 = (1+1)/(0+1) = 1/1 = 1. Note that we have the intuitive results $D + \overline{D} = 1$ and $D \cdot \overline{D} = 0$. As an example, the truth table of an AND operation on this 5-valued system is given in Table 5.2.

Table 5.2. The 5-valued AND operation in the D-algorithm.

	D	$\overline{D}$	0	1	x
D	D	0	0	D	x
$\overline{D}$	0	$\overline{D}$	0	$\overline{D}$	x
0	0	0	0	0	0
1	D	$\overline{D}$	0	1	x
x	x	x	0	x	x

To guide the decision making, the D-algorithm mains two groups of gates, the *D-frontier* and the *J-frontier* throughout the search process. A gate is in the D-frontier iff its output is unassigned and at least one of its inputs is assigned to D or $\overline{D}$. A gate is in the J-frontier iff its output is assigned but not yet justified by its inputs.

A *forward propagation* is simply the 5-valued simulation of assignments made in the circuit; a *backward implication* is a required assignment to the inputs of a gate because of an earlier assignment to its output. For example, if the output of an OR gate is 0, then both the inputs must be 0, whereas an implication can be made if the output is 1. The *controlling value* of a gate is an input assignment that determines the output regardless of the assignments to the other inputs. For example, the controlling value for an AND gate is 0. The non-controlling value is the negation of the controlling value. A conflict occurs if an assigned line is assigned again to a different value (except for assigning to x, which happens only in backtracking).

Based on the above definitions, the D-algorithm proceeds as follows. Note that the D- and J-frontiers are implicitly updated whenever a value change has occurred.

- Initially, the line at which the target fault is inserted is assigned to D if the fault is stuck-at-0 or to $\overline{D}$ otherwise. All other lines are assigned to x.

- After the initialization, the following steps are performed repeatedly. The algorithm terminates with a success if a D or $\overline{D}$ value is assigned to an output of the circuit and the J-frontier is empty, or with a failure if the entire search space is explored before a success is reached.

  - A decision is made to justify the value of a J-frontier gate output or to propagate the $D/\overline{D}$ values at a J-frontier gate input. In the former, one unassigned input is assigned such that the output is justified; in the latter, all unassigned inputs are assigned to the non-controlling value.

  - Forward propagation and backward implication are carried out iteratively, until no more propagation and implication are possible, or until a conflict occurs.

  - In case of a conflict, a backtracking is invoked to cancel the decision just made.

5.7.2 The PODEM and FAN algorithms

PODEM improves the D-algorithm by making decisions only on the inputs of the circuits. This is based on the observation that all internal assignments can be obtained by assigning only to the inputs. As a result, the search space in PODEM is exponential only in the number of inputs, as opposed to in the number of lines (gates) in the D-algorithm.

Deriving an input assignment from an internal line justification requirement is done via the *backtrace* process. Suppose a value v needs to be justified at line l. The backtrace process will find a path from l back to an input k and an assignment u to k, such that all the lines on the path are unassigned, and $u = v \oplus p$ with p being the inversion parity of the path. The intuition is that this assignment at the input k is likely to justify the assignment at l.

The PODEM algorithm proceeds by resolving line justifications only as the result of propagating the $D/\overline{D}$ values in the D-frontier. Therefore, the J-frontier is eliminated. The algorithm is shown in Figure 5.14.

The FAN algorithm is further improvement over the PODEM algorithm in many ways, including

```
podem() {
    if (D or D̄ reaches an output) return Success;
    if (D-frontier is empty) return Failure;
    if (conflict occurs) return Failure;
    (l, u) = an input assignment to a D-frontier gate;
    (k, v) = backtrace(l, u);
    propagate(k, v);
    if podem() == Success) return Success
    propagate(k, v̄);
    if podem() == Success) return Success
    propagate(k, v);
    return Failure;
}
```

Figure 5.14. The PODEM algorithm.

- more efficient backtracking that stops at heuristically selected internal lines,

- multiple simultaneous backtracings for the justification of a set of objectives, and

- immediate assignment to uniquely-determined lines to prune the search space.

As a result, when compared to PODEM, FAN reduces the number of backtracks and shortens the time between backtracks.

Chapter 6

CONSTRAINED VECTOR GENERATION

An important task in simulation is the generation of simulation vectors. This task can be time-consuming and error-prone, especially in the presence of complex interactions between the design and its environment. Vector generation methods fall into two major categories, the deterministic and the random-based. The former refers to the generation of vectors either manually, by engineers with a detailed understanding of the design, or automatically, by programs using ATPG techniques. This leaves the method highly sensitive to the complexity of the design. In practice, a design is usually first simulated with a relatively small set of deterministically created vectors, then with random vectors for as long as is feasible. Therefore, random vector generation is of great importance to simulation-based verification.

In this chapter, we describe a constrained random vector generation tool, called *Simgen*, that is based on symbolic techniques. The first requirement for any vector generation tool is to define, and to generate within, the legal input space that is often a function of the state of the design. In addition, a robust tool should also provide the ability to influence the distribution of the generation, directing the simulation towards the "corner" cases or other test scenarios that are deemed interesting. These two requirements can be simultaneously satisfied by representing and solving Boolean constraints and input biases using annotated BDDs. The vector generation algorithm consists of two procedures: the first computes the weights of vectors, composed from constraints and biases, and the second generates vectors according to probabilities derived from the weights. Both procedures operate symbolically using BDDs. The algorithm is immune to the *backtracking* problem that is typical to many search-based tools. The algorithm is also immune to the variable ordering problem despite the use of BDDs.

6.1   Constraints and Biasing

As we stated in Chapter 2, constraints are formal and unambiguous specifications of design behaviors and aspects. In the context of constrained random
simulation, there are basically two types of constraints: the environment constraints and constraints used as test directives. The environment constraints
define the interface protocol that must be strictly followed. The test directives take the form of the biasing of variables. Variable biasing will steer the
simulation to desired test scenarios, or the so-called "corner cases." For test
generation purposes, we do not distinguish these two types of constraints. We
treat all constraints as a whole which define the *legal input space*.

Biases or desired input distributions influence the randomization in test generation. They do not shape the legal input space. Instead, they are a means for
choosing solutions from the legal input space.

6.1.1   BDD Representation of Constraints

Constraints are represented implicitly by BDDs. Unless otherwise stated, we
shall refer to *constraint BDD* as the *conjunction* of the BDDs of all constraints.
Recall that the BDD as a representation of a Boolean function is also implicitly
a representation of the onset of that function. The legal input space defined by a
constraint is captured by the set of paths in the corresponding BDD that lead to
the terminal node *ONE*. Each of these paths can be viewed as an assignment to
the variables on that path. The assignment to state variables (a *cube*) represents
a set of states, whereas the input assignment represents a set of input vectors
that are legal under each of these states.

The above derivation of the legal input space for a given state is effectively a
computation of the constraint BDD's *cofactor* relative to that state. Depending
on the constraint, the legal input space can be empty under certain states, referred
to as the *illegal states*, or, more intuitively, as the dead-end states, since the
simulation cannot proceed upon entering those states.

EXAMPLE 6.1   Consider the constraint $(s_1 + s_2 + x_1 + x_2 \leq 1)$, where s_1 and
s_2 are state variables, and x_1 and x_2 are inputs: the state $(s_1 = 1, s_2 = 1)$ is an
illegal state, since no assignments to x_1 and x_2 can satisfy the constraint; states
$(s_1 = 1, s_2 = 0)$ and $(s_1 = 0, s_2 = 1)$ have one legal input vector, $(x_1 = 0, x_2 = 0)$; and
the legal input vectors for state $(s_1 = 0, s_2 = 0)$ are the ones satisfying $(x_1 + x_2 \leq 1)$.
$\square$

We shall introduce other BDD notations to facilitate further discussion. For
convenience, we will not distinguish a constraint and its BDD representation in
the sequel. We use $r(f)$ to denote the root node of a BDD f, $v(\sigma)$ to denote the
variable corresponding to the node σ, and $t(\sigma)$, $e(\sigma)$ to denote the *then* and *else*
nodes of σ, respectively. By our convention, X stands for the input variables,

and Y stands for the state variables. They will appear in the following definitions and discussions without further declaration.

6.1.2 Input Biasing and Vector Distribution

It is often the case that in order to exercise "interesting" design scenarios, one needs to "bias" the vector generation within the legal input space. The biasing mechanism is based on the idea of *input probabilities*.

DEFINITION 6.1 The *input probability* of $x = 1$ is a function of the state with a range in $(0,1)$, denoted by $p^x(Y)$; the *input probability* of $x = 0$ is the function $1 - p^x(Y)$, denoted by $p^{\overline{x}}(Y)$.

The values 0 and 1 are excluded from the possible values of input probability since they impose "hard" constraints and should be expressed as what they are, i.e., by the constraints $x == 0$ and $x == 1$, respectively. An input probability can be given either as a constant, or as a function, as shown in the example below.

EXAMPLE 6.2 The following function (in Verilog [CAD] syntax) gives an example of setting an input probability to 0.2 if st is UP, to 0.8 if st is $DOWN$, and to 0.5 otherwise.

$$st == UP ? 0.2 : st == DOWN ? 0.8 : 0.5$$

This function can be evaluated directly by the simulator, e.g., through a PLI call in Verilog. The result is then used by the test generator for picking the next input assignment. □

Input probabilities are "soft" restrictions since they can be "reshaped" by the constraints, which assume higher priority. In extreme cases, the constraints can prohibit an input variable from taking a specific value at all times, even if the input variable is assigned a high probability for doing so.

6.1.3 Constrained Probabilities

Constraints and input probabilities can be conceptually unified by the *constrained probability* of input vectors. First, we define an auxiliary term, the *weight of an input vector*.

DEFINITION 6.2 Let $\alpha = \alpha_1 \alpha_2 \cdots \alpha_n$ be a vector of input variables $x_1, \ldots, x_n$. The *weight* of α, denoted by $\pi(\alpha, Y)$, is given by

$$\prod_{i=1}^{n} [\alpha_i \cdot p^{x_i}(Y) + (1 - \alpha_i) \cdot p^{\overline{x_i}}(Y)] \tag{6.1}$$

The constrained probability is then defined in the following:

DEFINITION 6.3 Let f be a constraint, s a state, and f_s the legal input space which is the cofactor of f with respect to s. The *constrained probability* of an input vector α, under s, is 0 if $\alpha \notin f_s$; otherwise, it is given by $\frac{\pi(\alpha, s)}{\sum_{\beta \in f_s} \pi(\beta, s)}$

Conceptually, the constrained probability of an input vector is the weight of that vector divided by the sum of the weights of all vectors that satisfy the constraint; the sum is zero if the given state is an illegal state.

6.1.4   An Example of Constrained Probability

Consider a design with four inputs, cmd[3], cmd[2], cmd[1], and cmd[0], and with the corresponding input probabilities 1/2, 1/3, 1/4, and 1/5. When there are no constraints, all vectors are possible and each has a probability that is the product of a set of input probabilities, as shown in Table 6.1. (Middle vectors are removed for brevity.) Note that the probabilities of all vectors add up to 1.

Table 6.1.   Example: explicit computation of vector probabilities.

1/2 cmd[3]	1/3 cmd[2]	1/4 cmd[1]	1/5 cmd[0]	probability of vector
0	0	0	0	$1/2 \cdot 2/3 \cdot 3/4 \cdot 4/5 = 24/120$
0	0	0	1	$1/2 \cdot 2/3 \cdot 3/4 \cdot 1/5 = 6/120$
0	0	1	0	$1/2 \cdot 2/3 \cdot 1/4 \cdot 4/5 = 8/120$
	$\vdots$			$\vdots$
1	1	0	1	$1/2 \cdot 1/3 \cdot 3/4 \cdot 1/5 = 3/120$
1	1	1	0	$1/2 \cdot 1/3 \cdot 1/4 \cdot 4/5 = 4/120$
1	1	1	1	$1/2 \cdot 1/3 \cdot 1/4 \cdot 1/5 = 1/120$
				$\sum = 120/120 = 1$

Now we add a constraint

$$\text{cmd[3:0]==4'b1000 || cmd[3:0]==4'b0100 ||}$$
$$\text{cmd[3:0]==4'b0010 || cmd[3:0]==4'b0001}$$

that restricts our choices to the four vectors shown, enforcing a "one-hot" property among the inputs. These vectors and their "probabilities" (actually, *weights*, since the sum is now less than 1) are given in Table 6.2.

Finally, the constrained probabilities are obtained by normalizing the weights with regard to the total weights of the legal vectors. The results are 3/25, 4/25, 6/25, and 12/25, respectively, for the vectors in Rows 1 through 4.

The drawback of the above tableau approach is obvious: the cost is in the order of 2^n for n inputs. Next, we give an algorithm that computes the constrained probabilities implicitly in BDDs.

Table 6.2.   Example: explicit computation of vector weights under constraints.

1/2 cmd[3]	1/3 cmd[2]	1/4 cmd[1]	1/5 cmd[0]	*unnormalized* *weight of vector*
0	0	0	1	$1/2 \cdot 2/3 \cdot 3/4 \cdot 1/5 = 6/120$
0	0	1	0	$1/2 \cdot 2/3 \cdot 1/4 \cdot 4/5 = 8/120$
0	1	0	0	$1/2 \cdot 1/3 \cdot 3/4 \cdot 4/5 = 12/120$
1	0	0	0	$1/2 \cdot 2/3 \cdot 3/4 \cdot 4/5 = 24/120$
				$\sum = 50/120 \neq 1$

6.2   Simulation Vector Generation

The implicit computation of constrained probabilities consists of two procedures, *Weight* and *Walk*. The algorithm proceeds as follows: *Weight* computes
the *weights* of the nodes in the constraint BDD for a given state; depending
on the result, the algorithm either terminates because the state is illegal, or the
algorithm continues to the *Walk* procedure to generate an input vector. *Walk*
traverses the constraint BDD according to *branching probabilities* derived from
the weights of nodes. This flow is illustrated in Figure 6.1. The traversed path,
together with random assignments to input variables absent from this path,
represents an input vector that holds the following properties:

PROPERTY 6.1 The vector is a legal vector.

PROPERTY 6.2 The vector can be any legal vectors.

PROPERTY 6.3 The vector is generated with its constrained probability, as
    given in Definition 6.3.

The first two properties are necessary for a sound and complete simulation
vector generation process that produces *only* and *all* the vectors that satisfy the
constraint. The third property provides a utility for controlling the distribution
of the generated vectors. Because the algorithm operates on a BDD annotated
with probabilities, we call it the *p-tree* algorithm.

Figure 6.1.   Two-stage vector generation – *Weight* and *Walk*.

```
Weight(σ, s) {
    if (σ == ONE) return 1;
    if (σ == ZERO) return 0;
    if (σ is visited) return σ.weight;
    set_visited(σ);
    let u be the variable v(σ);
    if (u is an input variable)
        wt = Weight(t(σ),s);
        we = Weight(e(σ),s);
    else if (u is a state variable)
        if (u == 1) σ.weight = Weight(t(σ),s);
        else σ.weight = Weight(e(σ),s);
        return σ.weight;
    σ.weight = pᵘ(s) · wt + pū(s) · we;
    return σ.weight;
}
```

Figure 6.2.   Procedure *Weight*.

6.2.1   The *Weight* Procedure

First, we define the *weight* of a BDD node under a particular state.

DEFINITION 6.4 Given a constraint and the set of state variables Y, the weight of node σ, denoted by $\omega(\sigma, Y)$, is inductively given by the rules:

- $\omega(ONE, Y) = 1, \omega(ZERO, Y) = 0$

- Let Y be the set of state variables, and v the variable corresponding to ω; let t and e be the *then* and *else* nodes of ω, respectively. Then

$$\omega(\sigma, Y) = \begin{cases} p^v(Y) \cdot \omega(t(\sigma), Y) + p^{\bar{v}}(Y) \cdot \omega(e(\sigma), Y) & \text{if } v \text{ is an input variable} \\ \omega(t(\sigma), Y) & \text{else if } v = 1 \\ \omega(e(\sigma), Y) & \text{else} \end{cases}$$

$$(6.2)$$

The *Weight* procedure, as shown in Figure 6.2, applies the above computation of node weights recursively to the constraint BDD in a depth-first order. The following notations are used: *node*.var represents the variable associated with a BDD *node*; *node*.then and *node*.else represent the child nodes of *node*, for the assignments *node*.var=1 and *node*.var=0, respectively.

Weight performs a one-pass computation of node weights through the constraint BDD. A straightforward upper bound on the time complexity of *Weight* is $O(n)$, for a constraint BDD with n nodes. Note, however, that the procedure traverses only a subgraph in the BDD because it explores only one branch of each state node encountered. Furthermore, the nodes in this subgraph with positive (nonzero) weights form yet another subgraph that identifies the current legal input space; as we have shown in Section 6.1.1, the latter subgraph is effectively the cofactor of the BDD with respect to the current state. This cofactoring is "in-place" in the sense that it never creates new BDD nodes, thus avoiding potential BDD size explosion of normal BDD cofactoring. For these reasons, *Weight* is fairly efficient in practice even when the constraint BDD is quite large.

The second function of *Weight* is to determine whether the current state is legal, i.e., allows some satisfying input assignment. If it is, then we continue with the vector generation process; otherwise, we have to abort the simulation and start debugging the constraints. The following theorem provides such a test based on the result of *Weight*. Recall that $r(f)$ returns the root node of the BDD of f.

THEOREM 6.5 Given a constraint f, a state s is a legal state iff $\omega(r(f), s) > 0$.

PROOF: Input probabilities are always greater than 0, as are the weights of any input vectors. Therefore, the existence of satisfying input vectors indicating that the state is legal is equivalent to the sum of weights of satisfying vectors being positive.

The theorem is then the immediate result of Lemma 6.6 below, which says that the weight of the root node of f is the sum of the weights of the satisfying input vectors under s in f. □

LEMMA 6.6 Given a constraint BDD f and a state s, $\omega(r(f), s) = \sum_{\alpha \in f_s} \pi(\alpha, s)$ where f_s is the set of legal vectors in f under the s.

PROOF: See Appendix. □

6.2.2 The *Walk* Procedure

If the current state is a legal state, we proceed to actually generate a vector. Intuitively, the generation procedure resembles the reverse of evaluating a BDD for a given vector. In test generation, branches are taken electively, and assignments are made accordingly, whereas in BDD evaluation, the direction of branching is forced by the assignments given in the vector. This means that the key to test generation is how to branch electively. The solution is to follow the *branching probabilities*, built up from the weights just computed, in the following way.

```
Walk(node,st) {
    if (node == ONE)  return;
    if (node == ZERO)  error;
    let v be the variable node.var;
    if (v is a state variable)
        if (v == 1) Walk(node.then);
        else Walk(node.else);
    else
        node.then_prob = p(v) · t / node.weight;
        let r = random(0,1);
        if (r < node.then_prob)
            assign node.var to 1;
            Walk(node.then);
        else
            assign node.var to 0;
            Walk(node.else);
}
```

Figure 6.3.   Procedure *Walk*.

DEFINITION 6.7 Let σ be an input node with a positive weight, and u the associated variable. Let s be the state. The *branching probabilities* of σ are given in the following equations:

$$\sigma.\text{then\_prob} = [p^u(s) \cdot \omega(t(\sigma), s)]/\omega(\sigma, s) \qquad (6.3)$$

$$\sigma.\text{else\_prob} = [p^{\bar{u}}(s) \cdot \omega(e(\sigma), s)]/\omega(\sigma, s) \qquad (6.4)$$

Note that

$$\sigma.\text{then\_prob} + \sigma.\text{else\_prob} = 1. \qquad (6.5)$$

We use the branching probabilities to guide a random traversal in the constraint BDD, and generate a vector as follows. The traversal starts from the root node. At a state node, it takes the *then* (resp. *else*) branch if the corresponding state variable evaluates to 1 (resp. 0) in the current state. At an input node, it takes a branch according to the *branching probabilities* of the node and sets the value of the corresponding input variable accordingly, i.e., to 1 if a *then* branch was taken, to 0 otherwise. This procedure is implemented in Figure 6.3, and is named *Walk*.

Two properties, however, need to be proven to validate the *Walk* procedure. First, all the nodes that can be visited must have positive weights, since these are the only nodes where the branching probabilities are defined. The following theorem states exactly this.

THEOREM 6.8 Under a legal state, the procedure *Walk* only visits nodes with positive weights.

PROOF: See Appendix. □

As a result of the above theorem, the last node of the traversal must also have a positive weight and must be the node *ONE*.

COROLLARY 6.9 Under a legal state, the procedure *Walk* terminates only at node *ONE*.

Since the constraint BDD is satisfied by assignments corresponding to any path from the root to node *ONE*, this corollary asserts the second property we wanted to show about *Walk*: it generates only the legal vectors. The corollary, from a different viewpoint, also proves our claim that *p-tree* does not backtrack: if any legal input vectors exist, the *p-tree* algorithm guarantees to find one in a single pass.

One important point is that, since the input assignments made during *Walk* have already satisfied the constraint, we are totally free to assign the variables not visited in the traversal. In our case, we choose to do so according to their input probabilities. The default 0.5 is assumed if an input probability is not specified.

Finally, regarding the complexity of the p-tree algorithm, *Walk* always terminates within k steps on BDDs with k variables, since that is the length of the longest path. Overall, since k is a much smaller number than the number of nodes, the time and space complexities of *Weight* and *Walk* combined are linear in the size of the constraint BDD.

6.2.3   Correctness and Properties

Recall that our goal was to generate vectors with the three properties given at the beginning of this section. We now show that the *Weight* and *Walk* procedures have achieved this goal.

The first property, which says that the *p-tree* generates only legal vectors, is guaranteed by Corollary 6.9. This is because the constraint BDD is satisfied by assignments corresponding to any path from the root to node *ONE*.

The second property, that the algorithm can generate all legal vectors, is satisfied by the third, which states that the generation follows the constrained probabilities, because, by virtue of Definition 6.3, the set of legal vectors is exactly the set of vectors with positive constrained probabilities. Therefore, if

the third property holds, each legal vector will have a greater than 0 chance to
be generated. We restate the third property in the following theorem.

THEOREM 6.10 Procedure *Walk* generates input vectors according to their
constrained probabilities.

PROOF: See Appendix.   □

Since constrained probabilities are completely determined by the constraint,
the current state, and the input probabilities, a direct result of the above theorem
is the *p-tree* algorithm's independence of the BDD variable ordering. This
independence is a nice property for techniques based on BDDs.

COROLLARY 6.11 The probability of generating an input vector using the
p-tree algorithm is independent of the variable ordering of the constraint BDD.

Finally, we show that *p-tree* holds another property that correlates the input
to the branching probabilities.

LEMMA 6.12 Using the p-tree algorithm, the probability of generating an
input vector in which an input variable x_k equals 1 (resp. 0) monotonically
increases as p^{x_k} (resp. $p^{\overline{x}_k}$) increases.

PROOF: See Appendix.   □

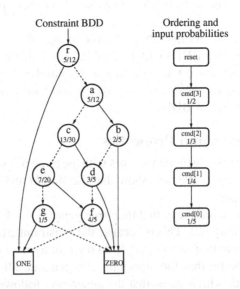

Figure 6.4.   A constraint BDD labeled with the node weight.

Table 6.3. Example: constrained probabilities (*reset=0*).

1/2 cmd[3]	1/3 cmd[2]	1/4 cmd[1]	1/5 cmd[0]	constrained probabilities
0	0	0	1	3/25
0	0	1	0	4/25
0	1	0	0	6/25
1	0	0	0	12/25

EXAMPLE 6.3 We modified the example from Section 6.1.4 in order to demonstrate the effect of state on constraints by qualifying the constraint with the state !*reset* (read *not reset*) as in the following:

> !reset→
> (cmd[3:0] == 4'b1000) ||
> (cmd[3:0] == 4'b0100) ||
> (cmd[3:0] == 4'b0010) ||
> (cmd[3:0] == 4'b0001).

The resulting constraint BDD is shown in Figure 6.4. Solid arcs represent *then* branches, dashed arcs represent *else* branches, and the variable ordering is shown to the right of the BDD. To aid the discussion, we use $x_0, \ldots, x_3$ to denote the inputs cmd[0], . . . ,cmd[3], respectively.

As described in Section 6.2.1, for a given state, only a subgraph in the BDD is involved in the computation. For the state "reset=1," the subgraph contains only the *ONE* node, meaning that all inputs are possible, and the constrained probability of a vector $\alpha = \alpha_0 \cdots \alpha_3$ is simply the product $\prod_{i=0}^{n}(\alpha_i \cdot p^{x_i} + (1 - \alpha_i) \cdot p^{\overline{x_i}})$. For the state "reset=0," the subgraph contains nodes a through g, which are the BDD corresponding to the original constraint as used in Section 6.1.4. We will show, under the state "reset=0," how *p-tree* implicitly computes the probabilities of legal inputs, which should match the constrained probabilities obtained in Section 6.1.4 and given again in Table 6.3.

First, *Weight* is applied to the root node r, which recurs in a depth-first order in the subgraph rooted at a. We illustrate the calculation by considering nodes f and e. For brevity, we denote the state "reset=0" by "!r".

$$\begin{aligned} w(f,!r) &= (1-p^{x_0}(!r)) \cdot w(ONE,!r) + p^{x_0}(!r) \cdot w(ZERO,!r) \\ &= 4/5 \cdot 1 + 1/5 \cdot 0 \\ &= 4/5 \end{aligned}$$

Similarly, we have $\omega(g, !r) = 1/5$. Then,

$$
\begin{aligned}
\omega(e, !r) &= p^{x_1}(!r) \cdot \omega(f, !r) + (1 - p^{x_1}(!r)) \cdot \omega(g, !r) \\
&= 1/4 \cdot 4/5 + 3/4 \cdot 1/5 \\
&= 7/20
\end{aligned}
$$

The weights of r and a through g are shown in Figure 6.4. The probabilities of legal input vectors are then obtained according to the procedure *Walk*. For example, vector 1000, which corresponds to the path r, a, b, d, f, ONE, has the probability $\frac{\frac{1}{2} \cdot \frac{2}{5}}{\frac{5}{12}} \cdot \frac{(1-\frac{1}{3}) \cdot \frac{3}{5}}{\frac{2}{5}} \cdot \frac{(1-\frac{1}{4}) \cdot \frac{4}{5}}{\frac{3}{5}} \cdot \frac{(1-\frac{1}{5}) \cdot 1}{\frac{4}{5}} = \frac{12}{25}$. Note that this probability matches the constrained probability of the same vector, shown in Table 6.3.   □

6.3   Implementation Issues

In this section we will discuss issues regarding the efficiency of the algorithm, with particular attention to the size of the constraint BDDs. We also describe an implementation of the algorithm in a commercial simulator.

6.3.1   Variable Ordering

As we have seen in the previous sections, the *p-tree* algorithm takes time and space linear in the size of the constraint BDD. Therefore, minimizing the BDD size is of a great interest to us.

BDD variable ordering must be mentioned here. We have shown in the introduction chapter that variable ordering can have a dramatic effect on BDD size, sometimes making an exponential-vs-linear complexity difference. Similar cases are abundant in hardware constraints. For example,

(st != 2'b11) ? (attr == Prev_attr) : 1.

If all the variables in "attr" occur before the ones in "Prev_attr," the equation would give a BDD with a size exponential to the width of "attr," whereas an interleaved ordering of the two would result in a linear size.

We developed heuristics that identify constructs such as those above to obtain a "good" initial order. In many cases, we observed that sticking to this initial order, rather than dynamically reordering the variables, renders faster execution of our algorithm.

6.3.2   Constraint Partitioning

When there is a large number of constraints, forming the conjunction BDD can be very expensive, for two reasons: (1) the computation blows up because of large intermediate BDDs; (2) the large conjunction BDD can slow down vector generation. *Simgen* partitions the constraint BDDs into groups with disjoint input variable support using the following procedure:

1. For each input variable, create a group.

2. For each constraint depending on a variable, add the constraint to the variable's group.

3. Merge all groups that share a common constraint until each constraint appears in at most one group.

4. Add all constraints that are not yet in any group to a new group. Observe that these constraints depend only on state variables.

The *p-tree* algorithm can then be applied to each group separately. The soundness of the *p-tree* under constraint partitioning is guaranteed by the following theorem:

THEOREM 6.13 Let C be a set of constraints, $C_1, \ldots, C_n$ be the disjoint-input-support partition of C, and $\alpha^1, \ldots, \alpha^n$ be the corresponding partial vectors generated by applying *p-tree* to the $C_1, \ldots, C_n$. Let α denote the concatenation $\alpha^1.\alpha^2 \ldots \alpha^n$. Then, under any state, the probability of generating α^1, $\alpha^2, \ldots$, and α^n is equal to the constrained probability of generating α from C.

PROOF: See Appendix. □

6.3.3 The Overall Flow

The *p-tree* algorithm is implemented as a library function interfaced to the Verilog simulator [CAD]. After an initialization sequence, the simulator calls *p-tree* at every clock cycle when it needs a new input vector, for example, right before the rising edge of the clock. At that time, the *p-tree* algorithm performs a sequence of tasks, as illustrated in Figure 6.5.

Note that procedure *Weight* is not always necessary in every simulation cycle. In particular, if all state variables occur before the input variables in the constraint BDD, and the input probabilities are constants, then it is only necessary to compute the weights once at the beginning of the simulation.

Figure 6.6 provides a high-level view of *Simgen*. The *Constraint Compiler* reads a Verilog [CAD] model annotated with constraints, and then extracts and compiles the constraints into BDDs that will be used by the *Vector Generator*. During simulation, the Verilog-XL simulator evaluates input probabilities for the current state at each clock cycle. The *Vector Generator* then samples the current state and generates an input vector based on the constraint BDDs.

6.4 Variable Solve Order

We have seen how vector distributions are controlled by the input biases through the constrained probabilities, and how constrained probabilities can be accurately realized in a symbolic algorithm. A critical property of this scheme is

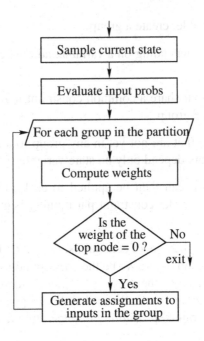

Figure 6.5. The *p-tree* algorithm.

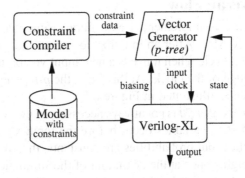

Figure 6.6. *Simgen* flow chart.

that when input biases are set to 0.5, as in the default case, all legal vectors are equally probable, that is, they assume a uniform distribution. This property is maintained during the course of simulation even if the set of legal vectors fluctuates with the change of the design state.

However, there are situations where a uniform distribution is not desirable. For example, for constraint $(c \rightarrow a == b)$ where a and b are 32-bit signals, the assignment $c = 1$ has a chance of only once in 2^{32} times. This shows that the uniform distribution on vectors can impose a very distorted distribution on variables. Input biases in *Simgen* are not probabilities but weights. They work collectively to control vector distribution, which ultimately determines the distribution of individual input variables. In a situation such as that above, [1] it is better to have more direct control at the variable level.

This can be accomplished by ordering variables in constraint solving. SystemVerilog Random Constraint (SVRC) provides the language support for this. In SVRC, the statement "solve *a* before *b*" declares that the variable *a* should be assigned before *b*. The problem in the previous example can be addressed with two additional ordering constraints, shown below in SVRC syntax.

```
constraint c1 (a -> b == c);
constraint c2 { solve a before b; }
constraint c3 { solve a before c; }
```

The last two constraints declare two *strict partial orders* $a > b$ and $a > c$. Thus *a* gets to pick its value first. Since both 0 and 1 are allowed for *a* in a solution, *a* is free to pick any value, meaning that an arbitrary probability can be assumed for *a*.

This idea of variable solve order can be implemented in *Simgen* as follows. First, we build a linear order from the set of strict partial orders. To do this, we construct a set of dags by applying the transitivity rule and merging the results. Note that no cycles are allowed, as per the SystemVerilog LRM [Acc03b]. Then we remove all the leaf nodes in the dags (last variables to solve in the orders). This is to minimize the number of ordered variables, since each of them requires extra effort in constraint solving. Finally, we coerce the dags into a compatible linear order by interleaving the nodes as much as possible. An example of the above process is shown in Figure 6.7.

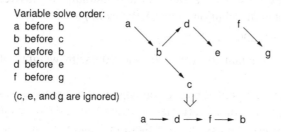

Figure 6.7. An example of variable solve order.

After the linear order is built, we preprocess the constraint BDD once for each ordered variable following this linear order. For each order variable v, if

---

[1] In this simple example, though, input biasing is good enough to give us a fair probability for variable c: just set the bias of $c = 0$ to $1/2^{32}$.

v can take both 1 and 0 in the BDD, we randomly pick a value for v, treat it as a state variable, and compute the node weights using the procedure, *Weight*. If v's value is fixed in the BDD, then nothing can be done and we proceed to the next variable. Whether or not a variable v can take both 1 and 0 can be easily determined in the weight computation of the previous variable. If any node of v has a *then* (resp. *else*) branch whose weight is not 0, we make a note that v can take the value 1 (resp. 0). For the first variable, before computing the weight, we do a plain BDD traversal to see if the variable can branch to any node other than the ZERO node on both the *then* and *else* branches.

Therefore, variable solve orders are handled by computing a linear order and performing a sequence of weight computations with a slightly modified *Weight* procedure, in order to decide if the next variable should be skipped.

It should be noted that because we do not force any variable assignment that is in conflict with the BDD (by skipping unassignable variables), the solution space remains the same. In particular, variable solve orders themselves will not cause the constraint solving to fail.

6.5   Weighted Distribution

Weighted distribution facilitates flexible randomization using input biasing and ordered variable solving. *Simgen* defines weighted distribution using the syntax

```
$dist {
    expr1 : weight1;
    expr2 : weight2;
    ...
    exprN : weightN;
}
```

where all the expressions are Boolean and mutually exclusive. SVRC also supports weighted distribution. The SVRC example

```
x dist { 100 := 1, 200 := 2, 300 := 5 }
```

specifies that x is equal to 100, 200, and 300 with weights of 1, 2, and 5, respectively.

Both can be implemented by introducing auxiliary random variables and by solving them ahead of other variables, as discussed in the previous section. For SVRC distributions, we first convert them into *Simgen*-like distributions. For example, the previous SVRC distribution would become

```
$dist {
    x==100 : 1;
    x==200 : 2;
    x==300 : 5;
}
```

As can be seen, the *Simgen* syntax is more robust than the SVRC syntax because it allows arbitrary Boolean expressions as opposed to equations in SVRC.

For each expression e_i, we create an auxiliary variable v_i and add the constraint $v_i == e_i$. For all the auxiliary variables of one distribution, we add a zero-one-hot constraint so that at most one of the expressions is asserted at any time. For example, we would have

```
constraint (v₁ == (x==100));
constraint (v₂ == (x==200));
constraint (v₃ == (x==300));
constraint (0 + v₁ + v₂ + v₃ < 1);
```

The probabilities of the auxiliary variables are the normalized weights of each expression (0.125, 0.25, 0.625, respectively, for the previous example). Finally, all of the auxiliary variables are ordered arbitrarily, and the constraint solving for ordered variables is applied so that these auxiliary variables are solved before other variables.

6.6 Cycling Variables and Permutations

A cycling variable cycles through all the values in a random permutation of the variable's range, as defined in SVRC (Section 3.3.6). Cycling variables can be implemented using weighted distribution. Let x be a variable with a finite range of size n. Let $(p_1, p_2, \ldots, p_n)$ be a permutation of that range. Let c be a counter that counts cyclically from 1 to n, and M an integer constant such that $M \gg n$. The following weighted distribution makes x a random variable cycling through the permutation.

```
$dist {
    x==p₁ : M*(c==1) + 1;
    x==p₂ : M*(c==2) + 1;
    ...
    x==pₙ : M*(c==n) + 1;
}
```

The number M is used to assign a large weight to the current number of choice. The role of those 1s is to prevent a 0 weight. Recall that in Section 6.1.2, no variable biases can take the value 0 or 1, which forces a hard constraint on the variable and may lead to failures in constraint solving. Therefore, a required assignment from a distribution can only be approximated with a very high weight. In fact, this restriction reflects the fundamental difference between a distribution and a constraint. A cycling iteration, for example, is not always guaranteed, because a constraint, especially a dynamically configurable one, may prohibit an assignment at the very moment when the iteration cycles to it. Constraints always take higher priority.

With a minor modification, a different permutation can be used for each iteration through the range. Simply replace the p_i's with variables and assign them a new permutation every time the counter counts to 1.

6.7 Historical Perspective

In [FDTM95], a tool called RIS implemented a static-biased random generation technique that allows the user to bias the simulation generation within a
restricted set of choices — all of which satisfy the constraints. However, the
biasing is static in that it is independent of the state of the design. To provide
more biasing flexibility, Aharon implemented a dynamically-biased test generator for a tool called RTPG [ABDD+90, ABDD+91], which decides the next
input based on the current state of the design. The primary drawback for the
simulation generator is the effort required to produce the functional model. A
tool introduced in [CI92] used various constraint solving techniques tailored for
specific instructions. A problem is that one may need to *backtrack* and perform
a heuristic search to resolve the "dead end" cases.

BDDs have found many applications in design and verification problems.
The Algebraic Decision Diagram (ADD) [BFG+93a], an extension to the BDD,
was used in [HMPS94] to represent the matrix of a state transition graph and to
compute the steady-state probabilities for Markov chains. Although simulation
generation was not the concern of that paper, experiments showed that symbolic
methods can handle very large systems that direct equation solving methods
cannot.

In [BCW80], binary graphs were used to represent Boolean functions, so that
the probability distribution of the output could be computed recursively from
the input probabilities of the function. This technique was used to probabilistically determine the equivalence of Boolean functions. A similar approach was
adopted in [KBS93] to compute exact fault detection probabilities for given
nets in a design.

Both [BCW80, KBS93] are related to our work by the idea that probabilities
are computed recursively using a decision diagram, but this is where the similarity ends. Specifically, we do not deal with the probability of the output of a
function or a design net. The problem we face is to generate input vectors that
both satisfy a set of constraints, and are probabilistically influenced by the input
distribution, or biasing. Constraints and biasing can both be state-dependent.

6.8 Results

We present experimental results from having applied *Simgen* to industrial designs. These designs came with environment constraints developed by engineers rather than by verification experts. We also report the results of experiments on building constraint BDDs for five designs in Section 6.8.1. In
Section 6.8.2, we give a case study using one of the designs.

The underlying BDD package is CUDD [S*et al.*], developed at the University
of Colorado. All experiments were conducted on a 233 MHz UltraSPARC-60
machine with 512 MB main memory.

Table 6.4. Statistics of designs.

name	total vars	vars in cons	num of cons
block1	76	26	13
block2	178	59	10
block3	1437	153	11
block4	446	175	33
block5	407	297	34

Table 6.5. Building the constraint BDD without partitioning.

example	time (sec)	peak nodes	mem (MB)	result nodes
block1	0.0	6312	0.5	54
block2	5.0	5110	3.6	119
block3	26.0	6132	9.3	774
block4	885.5	303534	35.5	110858
block5	727.7	181243	25.0	82405

6.8.1 Constraint BDDs

Table 6.4 reports the complexity of our benchmark designs. The BDD variables reported include inputs and latches. Columns 2 through 4 give the total number of BDD variables, the number of variables used in the constraints, and the number of constraints, respectively. Note that *block5* has about 300 variables in its constraints.

The results of building constraint BDDs are shown in Tables 6.5 and 6.6, respectively for cases without and with constraint partitioning. A sifting-based variable reordering [S*et al.*] was enabled in all experiments with the same setting.

Without partitioning, *block4* and *block5* each have close to 100k nodes in the resulting BDD, as shown in Table 6.5. The number of peak intermediate BDD nodes grows to over 300k for *block4*.

Table 6.6 shows the effectiveness of using partitioning. Although the number of constraints among the designs varies from 10 to 34 among designs, the average constraints per partition is about 3, which is fairly small. As a result, many BDD conjunction operations are avoided, and the total BDD size is reduced. Partitioning gives *block1, 2* and *3* some modest improvement but reduces both time and space complexity for *block5* and *6* dramatically. The complexity of the

Table 6.6.   Building the constraint BDD with partitioning.

example	time (sec)	#pnodes	mem(MB)	#rnodes	#cons	#parts
block1	0.0	1022	0.5	43	13	5
block2	4.0	5110	3.4	103	10	7
block3	20.3	6132	8.4	609	11	9
block4	38.0	13286	6.3	1595	33	10
block5	33.4	22484	7.4	1962	34	9

designs (over 1000 variables) and constraints (close to 300 variables), together with the size of the constraint BDDs (less than 2000 nodes), demonstrates that our technique is feasible for medium, and even large designs.

6.8.2   A Case Study

Simgen was used to construct verification environments for several design blocks. In the following, we present its application to a $PowerPC^{TM}$ slave block, namely, *block5* from the previous subsection.

Our first task was to develop the environment constraints according to a specification for the block (including its interface) written in English. Because of the sequential nature of the specification, we needed to introduce auxiliary variables to remember previous input or state values, and to construct abstract state machines of the design. The use of auxiliary variables further strengthens the expressiveness of *Simgen*.

Conflicts between constraints often occur because of the high complexity and the large number of constraints involved. One to resolve constraint conflicts is to used *guarded* constraints (see Section 3.2.4 in Chapter 3). like the following:

$$st \mathrel{!=} 2'b11 \quad ? \quad a == Prev\_a :$$
$$(!b \mathbin{\&} !c) \quad ? \quad u == Prev\_u :$$
$$(!b \mathbin{\&} c \mathbin{\&} !(e == 0 \mathbin{\|} e == 4)) \quad ? \quad ((f < 7) \mathbin{\&} (f > 0)) :$$
$$(t == 0 \mathbin{\|} t == 4 \mathbin{\|} t == 7) \quad ? \quad !(z == 6 \mathbin{\|} z == 5) : 1.$$

Note that $Prev\_a$ and $Prev\_u$ are auxiliary variables holding the values of a and u on the previous clock cycle. Because of the comparator "$a == Prev\_a$," an interleaved ordering of bits of a and $Prev\_a$ should be used.

The constraints for this block were written in about two person-days. The end result was a concise specification of the environment in a 200-line Verilog file, including 34 constraints, the auxiliary variables, and abstract state machines. The benefit of constraining cannot be overemphasized. Unconstrained random

Table 6.7. Result of biasing.

biasing	idle states	active states	total (sec)
none	2977	14	6.8
static	2179	811	6.2
dynamic	1073	1918	6.3

simulations generally produce false negative results. We also noticed that, in unconstrained simulations, the X value was constantly generated on tri-state buses, indicating bus contentions, making the simulations meaningless.

Developing input biasing was mostly straightforward. For instance, we wanted to limit the frequency of external errors when testing the essential functionality of a design. This was expressed as

$$\text{\$setprob1(error, 0.2);}$$

There were also cases where we needed to consider more complex dynamic biasing. For example, even after we statically biased over a dozen critical input signals, three major state machines mostly stayed idle through trial simulations. After studying the design for about an hour (it should take the original designer much less time), we were able to find a set of dynamic input probabilities that stimulated many more events in active states, such as *read* and *write*. The following biasing instructs *Simgen* to give a much higher input probability to u_i in state *IDLE*, which triggers a transition out of *IDLE*:

$$\text{\$setprob1}(u_i, \text{addr\_state == IDLE ? 0.9 : 0.5);}$$

We wrote a few dynamic biasing commands for each of the state machines. Note that we did not use constraints to express the condition, because "leaving" the idle state is not a mandatory action, it simply serves our test purpose. Additionally, using input biasing instead of constraints can avoid potential constraint contradictions, because the constraints take priority over input probabilities.

Of course, in general, state transitions are controlled by both state and inputs, and justifying an enable condition can be a very hard problem. However, with the help of automatic tools such as *Simgen*, this problem can be minimized. Table 6.7 shows the effect of biasing on the number of active states (of three major state machines) visited during simulations of 1000 cycles each. The dynamically biased simulation increased the coverage 130 times over that of the unbiased simulation.

Simulations usually run with monitoring processes, or they dump the results to log files for post-processing. Table 6.8 summarizes the run-time overhead

Table 6.8.   Overhead of *Simgen*.

setting	overall (sec)	Simgen overhead
random	44.9	–
Simgen	48.2	21.3%
w. dump	63.6	16.0%
w. monitor	635.6	1.7%

of *Simgen* on *block5*, with partitioned constraint BDDs and dynamic biasing. All simulations ran for 10,000 cycles each. Rows 2 to 5, respectively, represent the simulations with pure random generation, stand-alone *Simgen*, *Simgen* with Verilog dump, and with property monitoring. It can be seen that the overhead of *Simgen* is fairly low.

In this specific example, *Simgen*, together with a simulation monitoring tool, discovered thirty design bugs, basically falling into two categories:

1. Bugs caught because the design entered a state where there was no possible legal input. This implies the design has violated the constraints.

2. Bugs caught because the design entered an illegal state. This is usually manifested as property violations.

6.9   Summary

We have described an automated simulation-vector generation method and tool. Constraints are used to generate legal vectors that are influenced by input biasing. Both constraints and biasing can depend upon the state of the design, providing robust environment modeling capability. The implementation, *Simgen*, is based on an efficient symbolic algorithm that does not backtrack in solving the constraints. The effectiveness of *Simgen* is demonstrated by its application to commercial design verification.

Before concluding this chapter, we would like to briefly mention experiments in extending the biasing method in *Simgen*, based upon the probabilities of individual inputs. We looked at a more general biasing given by the *event probabilities* of the form $Pr(Event) = \pi$, where *Event* is a characteristic function defining a set, for example, $In\_a == In\_b$. Unfortunately, Koller and Megiddo [KM94] showed that the problem of deciding whether there exists any distribution satisfying a set of event probabilities is NP-hard. We described, in [YA00], a decision procedure using Linear Programming (LP) that gives a distribution, if it exists, with a minimum number of positive LP variables. However, the number of variables in the worst case can be exponential to the number of event probabilities. In [YAA02] we reported a method of directed simulation using event probabilities without actually solving for them. In that case, we extracted a finite state machine for each desired property, and treated each state transition as an event. The probability of each transition is dynamically assigned depending on the current state, i.e., the transitions that are most likely to lead to the "bad" state in the finite state machine are given high probabilities. Experimental results showed that when a transition is possible, *Simgen*, with the new biasing method, has a very high probability of generating a vector that enables the transition.

Chapter 7

CONSTRAINT SIMPLIFICATION

An important difference between a hardware design constraint and a normal Boolean constraint is that the former is defined over two type of variables — the state variables and the inputs. Each state variable stores the value of a latch, which is determined by the driving logic of that latch, and the valuation of latches and inputs from the previous clock cycle (an exception is that the initial state is determined by the reset logic). Since they are pre-determined by the design, state variables are really parameters in hardware design constraints, whereas the inputs are the real variables we are solving for. This heterogeneousness of state and input variables has already been exemplified in the vector generation algorithm given in Chapter 6. In this chapter, we will use it for constraint simplification. In Chapter 9, we will revisit it for optimization of a different vector generation method.

Recall that all constraints with common input variables need to be conjoined before solving (disjoint-input partitioning), which can lead to performance degradation when there are many constraints to conjoin. Here we introduce a special class of constraints that is exempted from this restriction — the *state-deterministic* constraints [YAAP04]. A state-deterministic constraint has the following traits:

- Under certain states, the input variables are a function of the state, i.e., they are fully determined.

- Under all other states, the constraint evaluates to true regardless of the inputs.

For brevity, we will call the state-deterministic constraints *det-constraints*. Since state variables are parameters, we can always plug their values into a det-constraint and either get a fixed assignment to the inputs or discharge the constraint altogether. Correspondingly, we take the following actions on a det-constraint:

- If the inputs are fixed, propagate them to the other constraints for simplification.

- Otherwise, discharge the det-constraint.

In either case, the action on a det-constraint is independent of any other constraint, even if they have common input variables. However, the converse is not true since the input assignment from a det-constraint does propagate to other constraints.[1] It follows that all det-constraints must be solved before all none-det-constraints.[2] Note that propagation of an input assignment removes the input itself and, often, other inputs as well. Therefore, the benefit is twofold: simplified constraints, and a finer disjoint-input partition.

EXAMPLE 7.1 Suppose we have two constraints, $\bar{a} + x$ and $x + y$, where a is a state variable and x, y are inputs. The first constraint is a det-constraint since, when $a = 1$, x is fixed at 1, and when $a = 0$, the constraint is immediately true. In the former case, plugging $ax = 1$ into $x + y$ removes y, thus y can assume any value. In the latter, the det-constraint is discharged, and the problem reduces to solving the remaining constraint. It is easy to see the solution:

a	x	y
1	1	-
0	1	-
0	-	1

A little more effort would be required if we were to solve $(\bar{a} + x)(x + y)$ together, although we would arrive at exactly the same solution.   □

In [YKAP02], det-constraints are called *hold-constraints*. This dichotomy arises from a frequently encountered scenario in hardware design constraints — under certain conditions, an input variable maintains its value from the previous clock cycle, or it is simply fixed as a constant.

EXAMPLE 7.2 Let us look at some real-world examples of hold-constraints.

> *gclk   ?   qpc_ack == prev1_qpc_ack : 1*
> *gclk   ?   qpc_state == prev1_qpc_state : 1*
> *gclk   ?   xdbarw_in == prev1_xdbarw_in : 1*
> *gclk   ?   xdbbrw_in == prev1_xdbbrw_in : 1*

In the first constraint, the clock signal *gclk* and signal *prev1_qpc_ack* are the state variables. The latter is a one-cycle delayed version of the input *qpc_ack*.

---

[1] Indeed, the effect of hold-constraint extraction and the ensuing simplification resembles that of the derivation and propagation of constants in SAT.
[2] It does not matter in which order the det-constraints are solved.

Other constraints are similarly defined. These four constraints state that under certain clock phases, the inputs should hold their value from the previous cycle.

$$data\_tag == prev1\_data\_tag\_A1$$

data_tag is an input and *prev1_data_tag_A1* is the delayed version of *data_tag_A1*, which is also an input. This constraint says that input *data_tag* should always follow *data_tag_A1* in the next cycle.

$$prev2\_data\_tag\_A1[0] \ ? \ dval\_A1 : 1$$

If bit-0 of *prev2_data_tag_A1* (two-cycle delayed version of *data_tag_A1*) is asserted, then the input *dval_A1* (data valid) is also asserted.

Figure 7.1 shows the waveforms of the above det-constraints. □

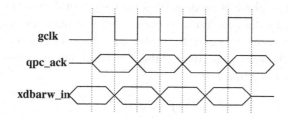

Figure 7.1. Waveforms for det-constraints.

Causal relationships between the input and state in terms of det-constraint are common in constraints used for testbench automation. Many of them are expressed in easily recognizable forms, as in the above examples. Because of the individualism in the style of constraints written by engineers, there are cases where a det-constraint may be tangled in a larger constraint, causing it to need to be *extracted*. In extreme situations, det-constraints can only be inferred when multiple constraints are combined. In this chapter, our task is to develop a robust and efficient method for extracting det-constraints and using them for simplifying the constraint solving problem.

7.1 Definitions

We recall that constraints are Boolean functions of input and state variables. The onset of a constraint represents the legal (state-dependent) input space regarding this constraint. The overall legal input space is the intersection of the onsets of all the constraints. We will use f, g, h, k, and e as function symbols, a, b, and c as state variables, x, y, and z as input variables, s_i, s_j as states, i.e., minterms of state variables, and S as a set of states.

DEFINITION 7.1 Let ω be the set of state variables. A *det-constraint* on input variable x is a constraint $e(x, \omega)$ such that there exist Boolean functions $k(\omega)$ and $g(\omega)$ such that $e(x, \omega) \leftrightarrow (k(\omega) \to (x = g(\omega)))$.

All det-constraints can be written in the following *normal* form:

$$k \to (x = g). \tag{7.1}$$

Note that since the context is clear, we drop the state argument ω from functions k and g. We call k the *condition* and g the *assignment* of the det-constraint. Note that both k and g can be constants, except that k cannot be 0.

We can infer a det-constraint $k \to (x = g)$ from f iff the following *implication test* is met:

$$f \to (k \to (x = g)) \tag{7.2}$$

However, this test alone is inadequate for a meaningful extraction because it can be met by any det-constraint whose condition k does not overlap with f. Such a det-constraint will never be "activated," since f has to be satisfied as a constraint, which means k is always false. To avoid this, we must also enforce the following *non-disjoint test*:

$$f \cdot k \neq 0 \tag{7.3}$$

DEFINITION 7.2 A det-constraint $k \to (x = g)$ is *extractable* from constraint f if the implication and non-disjoint tests are satisfied by k and f.

EXAMPLE 7.3 Figure 7.2 shows a real-world constraint from a data cache design. The constraint is written in Verilog-XL [CAD] where the symbols $\|$, &, and ! stand for the logic operations OR, AND, NOT, respectively.

This constraint relates the cache controller's internal state to the two inputs dc_freeze and $dc\_unfreeze$, which, respectively, disables and enables the transactions in the cache. The state variables are: $gclk$, the clock; mem_access, the flag that the cache engages in memory transactions; $addr\_frozen$ and $queue\_frozen$, the signals for deactivation of the address bus and the transaction queue, respectively; and $queue\_top$, the type of the top entry in the transaction queue, which can take "'dc_trans" (data cache transaction) among other values.

This will be our running example throughout the chapter. For ease of exposition, we abbreviate the signals dc_freeze, $dc\_unfreeze$, $gclk$, mem_access, $addr\_frozen$, $queue\_frozen$, and the predicate $queue\_top == {}'dc\_trans$ with x, y, a, b, c, d, and p, respectively. We also replace the Verilog-XL operators (except "?") with our own. We then have

$$a \ ? \ (((b+c) \ ? \ x \ : \ ((d \cdot \overline{y}) \ ? \ x \ : \ ((y+p) \ ? \ \overline{x} \ : \ x)))) \ : \ 1. \tag{7.4}$$

```
gclk?
    ((mem_access || addr_frozen)?
        dc_freeze :
        ((queue_frozen&!dc_unfreeze)?
            dc_freeze :
            ((dc_unfreeze || queue_top == 'dc_trans)?
                !dc_freeze :
                dc_freeze
            )
        )
    ) : 1
```

Figure 7.2. A constraint example.

Noting that $a?b : c$ means $(a \rightarrow b) \cdot (\bar{a} \rightarrow c)$, we infer from (7.4) the following:

$$a \cdot (b + c) \rightarrow x. \tag{7.5}$$

Furthermore, the above is a det-constraint, since the condition $a \cdot (b + c)$ involves only state variables, and the input x is fixed to a constant if the condition is true. The following det-constraint can also be inferred:

$$a \cdot \overline{b + c} \cdot \bar{d} \cdot p \rightarrow \bar{x}. \tag{7.6}$$

However, constraints can get much more complex than (7.4). For example, one may encode complicated conditions by using functions which in turn are used to write constraints. A constraint file with these intermediate function definitions can easily exceeds hundreds or even thousands of lines of code. This complexity has prompted the search for an automatic det-constraint extraction method. In the coming sections, we explore how such an extraction can be done both syntactically and functionally. □

7.2 Syntactical Extraction

Syntactical extraction is based on conjunctive and disjunctive decompositions using simple syntactical rules. Functions $g_1, ..., g_n$ define a conjunctive (resp. disjunctive) decomposition of function f if $f = g_1 \cdot ... \cdot g_n$ (resp. $f = g_1 + ... + g_n$). First, we try to decompose each constraint into a set of conjuncts. If a constraint is not conjunctively decomposible, the set contains just the original constraint. Then we examine these conjuncts to see if they can be disjunctively decomposed and, furthermore, if the disjunctions match any det-constraint in the normal form.

EXAMPLE 7.4 Suppose we have the following rewrite rules for conjunctive decomposition:

$$u \, ? \, v : w \; := \; (u \to v) \cdot (\overline{u} \to w) \tag{7.7}$$

$$u \to (v \to w) \; := \; u \cdot v \to w \tag{7.8}$$

Repeatedly applying Rule (7.7) to constraint (7.4), we have the following as one of the conjuncts

$$a \to ((b + c) \to x), \tag{7.9}$$

to which we apply Rule (7.8) and get

$$a \cdot (b + c) \; \to \; x. \tag{7.10}$$

which is exactly the function given in (7.5) previously. This function is a disjunction and further a det-constraint because the left hand side of "$\to$" is a condition depending only on the state, and the right hand side assigns an input to a constant.   □

Synt_extract {
    Conjunctively decompose the constraint according to
      a set of syntactical rules
    For each conjunct f
      If f cannot be decomposed into a disjunction $k + h$
        where k and h are Boolean functions
      Then skip f
      If k depends on input variables
      Then swap k and h
      If k depends on input variables
      Then skip f
      If h is in the form of an assignment $x = g$
        where x is an input variable and
        g does not depend on input variables (can be a constant)
      Then add $\overline{k} \to (x = g)$ to the det-constraint set
}

Figure 7.3. Syntactical extraction.

The syntactical extraction of det-constraints is summed up in the algorithm given in Figure 7.3. The algorithm satisfies the implication and non-disjoint requirements. Its time complexity is linear in the size of the constraint formulas, but exponential in the number of syntactical rules. Because of the latter, it is only practical to use a small set of simple rules. This hinders our ability to extract complex det-constraints. For example, we cannot obtain the det-constraint in (7.6) by using only Rules (7.7) and (7.8) – an inference rule "$((u + v) \to w) :\to (u \to w)$" has to be added. Unfortunately, even if we can afford to use an arbitrarily large set of rules, syntactical extraction is still incomplete because there are infinitely many ways to write constraints that are functionally identical. Next, we will present a functional det-constraint extraction algorithm that is both complete and efficient.

7.3 Functional Extraction

DEFINITION 7.3 Given a constraint f, an input x is *positively* (resp. *negatively*) *bounded* by a set of states S if, in all minterms in the non-empty intersection $f^{on} \cap S$, it is the case that x evaluates to 1 (resp. 0).

Intuitively, x is *positively bounded* in f by S if we can infer the implication $S \to x$ from f, and x is *negatively bounded* by S if we can infer the implication $S \to \bar{x}$ from f. Note that the implications are det-constraints and that f intersects S. Hence the det-constraint $S \to \bar{x}$(or $\bar{x}$) passes both the implication and non-disjoint tests and is extractable.

LEMMA 7.4 If input variable x is bounded by state set S in constraint f, then the det-constraint

$$S \to (x = b_i) \tag{7.11}$$

where b_i is 1 if x is positively bounded and 0 otherwise, is *extractable* from f.

EXAMPLE 7.5 Consider the constraint represented by its onset minterms:

a	b	x	y	z
0	0	1	-	-
0	1	1	1	-
1	0	-	0	-
1	1	1	-	0
1	1	0	-	1

where a and b are the state variables. Input x is positively bounded by the state set $\{(0,0),(0,1)\}$; y is positively bounded by $\{(0,1)\}$ and negatively bounded by $\{(1,0)\}$; z is not bounded by any state. Correspondingly, three det-constraints can be extracted: $\bar{a} \to x$, $\bar{a}b \to y$, and $a\bar{b} \to \bar{y}$. ☐

A systematic det-constraint extraction from a constraint is based on testing the boundedness of an input variable on all possible states, then combining all det-constraints detected this way to form a final det-constraint that encompasses all the states that bound the input. Hence, the final det-constraint is *maximal* with respect to the constraint being tested. However, two problems need to be addressed for automating the extraction. First, a simple-minded state-by-state testing, as shown in the above example, does not work when the state space is too large to enumerate. Consider, for example, listing the minterm form of the constraint in Figure 7.2 that has 5 state variables, which is still modest in terms of state space. Therefore, an implicit approach is in order. Second, the technique used to combine these det-constraints remains a question. It turns out that both problems can be solved by a fully symbolic procedure that evolves from a concept called the *prime* det-constraints.

DEFINITION 7.5 A det-constraint $k \to (x = g)$ is *prime* if g is a constant, and the onset of k is a singleton, i.e., contains exactly one state.

The theorem below indicates that the conjunction of the prime det-constraints obtained as in Lemma 7.4 on an input variable is a det-constraint on the same variable, and that the converse is also true.

THEOREM 7.6 The conjunction of a set of prime det-constraints on x with mutually exclusive conditions $s_1, \ldots, s_l$ is a det-constraint, and vice versa. That is,

$$\bigwedge_{i=1}^{l} \left(s_i \to (x = b_i) \right) \leftrightarrow k \to (x = g) \tag{7.12}$$

where $s_i \wedge s_j = 0$ for $i \neq j$, $b_i \in \{0, 1\}$; k and g can be derived from the prime det-constraints as

$$k = \bigvee_{i=1}^{l} s_i, \; g \in [g^{on}, \overline{g^{off}}], \; g^{on} = \bigvee_{i=1}^{l} (b_i \cdot s_i), \; g^{off} = \bigvee_{i=1}^{l} (\overline{b_i} \cdot s_i), \tag{7.13}$$

and the prime det-constraints can be derived from k and g as in the set

$$\{(s_i, b_i) \mid 1 \leq i \leq |k|, \; s_i \in k, \; b_i = g_{s_i}\}. \tag{7.14}$$

where $|k|$ is the number of minterms in the onset of k.

PROOF: See Appendix.   □

Because f implies and intersects with the condition of every prime det-constraint on x as obtained in Lemma 7.4, it must also imply the conjunction of the said prime det-constraint and intersect with the union of the said conditions. Therefore, if all of the prime det-constraints are extractable from f, the det-constraint conjunctively constructed in Theorem 7.6 is also extractable from f.

Now we construct the procedure that actually "computes" the construction in Theorem 7.6. Denote the set of input variables in $\chi - \{x\}$ by x'. The following is true for any constraint f:

1. $(\exists_{x'} f)_x - (\exists_{x'} f)_{\bar{x}}$ is the set of all the states with respect to which x is positively bounded.

2. $(\exists_{x'} f)_{\bar{x}} - (\exists_{x'} f)_x$ is the set of all the states with respect to which x is negatively bounded.

The union of the above two disjoint state sets is sometimes called the *Boolean differential* of $\exists_{x'} f$ with respect to x, denoted as $\partial(\exists_{x'} f)/\partial x$. The conjunction of the prime det-constraints conditioned on this set is the *complete* det-constraint on x, since it includes all the states for which x is bounded. The derivation of such a det-constraint is formalized in the theorem below, which follows naturally from Theorem 7.6 and the above analysis.

THEOREM 7.7 The complete det-constraint of f on x is

$$k \rightarrow x = [g^{on}, \overline{g^{off}}] \tag{7.15}$$

where

$$k = \frac{\partial(\exists_{x'} f)}{\partial x}, \ g^{on} = k \cdot (\exists_{x'} f)_x, \ g^{off} = k \cdot (\exists_{x'} f)_{\bar{x}}. \tag{7.16}$$

Obviously, this det-constraint meets the implication and non-disjoint requirements iff

$$\frac{\partial(\exists_{x'} f)}{\partial x} \neq 0. \tag{7.17}$$

This is the main theorem about functional det-constraint extraction. We can choose a function g from the interval $[g^{on}, \overline{g^{off}}]$ by using any *care-set optimization* function ψ, such that $g = \psi(g^{on}, c)$ and $\psi(g^{on}, c) \cdot c = g^{on} \cdot c$, where the care set $c = g^{on} + g^{off} = k$.[3] The resulting extraction is unique with respect to x and f up to the selection of ψ. We choose the BDD *Restrict* function [CM90, SHSVB94] since it is efficient, usually reduces the BDD size, and does not introduce new variables from the care set to the result.

Although the extraction is complete, the BDD representation of k can still be optimized with regard to f using BDD *Restrict*. Since $k \leq \exists_X f$, and due to properties of BDD *Restrict*, the onset of k only increases in the optimization, and the increment only comes from $\bar{f}$. Thus, the only side effect is the addition of "vacuous" prime det-constraints whose conditions are disjoint with f, which does not destroy the completeness property of the extraction.

EXAMPLE 7.6 Returning to the example given in (7.4), we show how Theorem 7.7 can be used to extract det-constraints for the inputs x and y. For brevity,

---

[3]Also note that the care set for $x = g$ is also k due to the det-constraint. However, since $k \leq \exists_X f$, function g does not need to be simplified with respect to f.

we omit the computation and only present the result. First, we expand the ? operation and rewrite (7.4) in the familiar AND, OR, and NOT notation that gives the following:

$$\overline{a} + (b + c + d \cdot \overline{y} + \overline{p} \cdot \overline{y}) \cdot x + \overline{b} \cdot \overline{c}(\overline{d} \cdot p + y) \cdot \overline{x}. \qquad (7.18)$$

Now, for input y, the computation of k results in 0. Therefore, no extraction is possible. For x, we have

$$
\begin{aligned}
k &= a \cdot (b + c + \overline{d} \cdot p) \\
g^{on} &= a \cdot (b + c) \\
\psi(g^{on}, k) &= b + c.
\end{aligned}
$$

The det-constraint for x is then

$$a \cdot (b + c + \overline{d} \cdot p) \;\to\; x = b + c. \qquad (7.19)$$

By simple Boolean algebraic argument, one can show that the above is exactly the conjunction of the det-constraints given in (7.5) and (7.6). Note that the latter was not detected by the syntactical extraction, as shown in Section 7.2.

□

7.4 Constraint Simplification

We have described how det-constraints can be extracted from constraints.[4] Now we explore how the extraction can be used to simplify the input constraints. We show that this can be done by applying the following *conditional substitution*.

DEFINITION 7.8 Let e be a det-constraint $k \to (x = g)$. The *conditional substitution* of e on a Boolean function f, written $\tau(f, e)$, is

$$\tau(f, e) = k \cdot f_{x:=g} + \overline{k} \cdot f \qquad (7.20)$$

where $f_{x:=g}$ is the substitution of variable x with function g.

An obvious advantage of conditional substitution is that, if a constraint is itself a det-constraint, then, after we extract that det-constraint and substitute it back, the original constraint will be reduced to 1[5] and removed from further

---

[4]The extraction in its own right helps speed up random vector generation for the following reason. For a det-constraint $k \to x = g$, both k and g are functions of state variables and thus can be easily evaluated once the design state is known. The det-constraint is discharged if k evaluates to 0. Otherwise x takes the value of g. This avoids the NP-hardness of a generic constraint solving problem.

[5]To see this, let $f = k \to x = g$. Then $\tau(f, f) = k \cdot (k \to g = g) + \overline{k} \cdot f = k + f = k + \overline{k} + (x = g) = 1$.

consideration. This effectively provides a mechanism to single out constraints that are det-constraints so that they can be solved separately.

In other cases, conditional substitution can remove the involved input variable from the constraint. For example, let $f = a + x + y$ and $e := \bar{a} \to \bar{x}$. Then $\tau(f,e) = \bar{a} \cdot (a+y) + a \cdot (a+x+y) = a+y$. If, in constraint partitioning, f were to be conjoined with other constraints because they share the input x, then using $\tau(f,e)$ instead of f would prevent this from happening. On the other hand, BDD *Restrict*, although widely used as a simplification function, is sensitive to variable ordering and may not improve the result. If x is the top variable, then restricting f on e returns f itself.

It turns out that the conditional substitution is a care-set optimization function, just like BDD *Restrict*, but insensitive to variable ordering. It also possesses other useful properties:

PROPERTY 7.1 $\tau(f,e) \cdot e = f \cdot e$, i.e., $\tau(f,e)$ is a function equivalent to f in the care set e.

PROPERTY 7.2 $\tau(f,e)$ decreases the "diversity" of x in f, i.e., $\partial f/\partial x$, by the amount k. This implies that $\tau(f,e)$ is independent of x iff $k \geq \partial f/\partial x$

PROPERTY 7.3 If $\tau(f,e)$ is independent of x and $f \to e$, then $\tau(f,e) = \exists_x f$.

PROPERTY 7.4 If there exists a care set optimization function $\psi(f,e)$ that does not depend on x, then it must be $\tau(f,e)$.

PROOF: See Appendix. □

Note that the last property makes conditional substitution a better choice than BDD *Restrict* with regard to input variable removal. Proofs of these properties are given in the appendix.

Although conditional substitution does not necessarily remove the constraint or any input variables in all cases, it generally simplifies the constraint as indicated by Property 1. We can confirm this by applying conditional substitution to constraint (7.18) (the rewrite of (7.4)) using its det-constraint (7.19). The result is

$$\bar{a} + b + c + \bar{d} \cdot p + \bar{y} \cdot x + y \cdot \bar{x}. \tag{7.21}$$

which has nine literals while (7.18) has fourteen. The fact that x is not removed after the substitution is predicted by Property 2. For constraint (7.18), $\partial f/\partial x = a$, and for det-constraint (7.19), $k = a \cdot (b+c+\bar{d} \cdot p)$. Therefore, $k < \partial f/\partial x$ is viable – according to Property 2, the substitution will depend on x.

7.4.1 Recursive Extraction

Extracted det-constraints can be used to extract more det-constraints than would otherwise be possible. When there are multiple constraints, more extractions can be done by conjoining the constraints and applying the procedure in Theorem 7.7. However, this approach is not practical, since conjoining the original constraints can be expensive. Therefore, after having extracted from individual constraints, by using Theorem 7.7, we again try only the conjunction of each constraint and the extracted det-constraints, whose size is usually small. In fact, we can even avoid explicitly conjoining a det-constraint with a constraint, due to the following theorem.

THEOREM 7.9 For any det-constraint, $e := k \rightarrow (x = g)$, and Boolean function f, an input variable $y \neq x$ is extractable from $\tau(f, e)$ iff it is extractable from $f \cdot e$, or more precisely,

$$\frac{\partial(\exists_{y'} \tau(f, e))}{\partial y} = \frac{\partial(\exists_{y'} (f \cdot e))}{\partial y} \tag{7.22}$$

where $y' = X - \{y\}$.

PROOF: See Appendix.   □

The above theorem implies that, given a constraint and a det-constraint, conditional substitution is an exact method for finding det-constraints for input variables other than the one being substituted.

EXAMPLE 7.7 We again use the example in (7.4) to illustrate the usefulness of this theorem. We have already shown that (7.4) is simplified to (7.21) using the det-constraint (7.19) extracted from (7.4). Now suppose that from some other constraint, we have a new extraction

$$\overline{b} \cdot \overline{c} \rightarrow \overline{y}. \tag{7.23}$$

Conditionally substitute this into (7.21), and we have

$$\overline{a} + b + c + \overline{d} \cdot p + x \tag{7.24}$$

from which we can extract yet another det-constraint

$$a \cdot \overline{b} \cdot \overline{c} \cdot (d + \overline{p}) \rightarrow x. \tag{7.25}$$

Furthermore, since (7.24) is the det-constraint (7.25), conditional substitution using (7.25) reduces (7.24) to 1. Therefore, with the help of the det-constraint (7.23) extracted elsewhere, constraint (7.4) is eventually removed and replaced by two det-constraints, (7.19) and (7.25), which can be solved easily and independently.

Note that the conjunction of det-constraint (7.23) with constraint (7.21) is a function that is more complex than (7.24) (involving both inputs x and y) but that yields the same det-constraint on x as (7.25), as predicted by Theorem 7.9.

$$\bar{a} + (b + c + d\bar{y} + \bar{p}\bar{y})\,x + \bar{b}\bar{c}\,(\bar{d}p + y)\,\bar{x} \quad \xrightarrow[\text{simplify}]{\text{extract}} \quad a(b + c + \bar{d}p) \to x = b + c$$

$$\Downarrow$$

$$\bar{a} + b + c + \bar{d}p + \bar{y}x + y\bar{x} \quad \xleftarrow{\text{simplify}} \quad \bar{b}\bar{c} \to \bar{y}$$

$$\Downarrow$$

$$\bar{a} + b + c + \bar{d}p + x \quad \xrightarrow[\text{simplify}]{\text{extract}} \quad a\bar{b}\bar{c}\,(d + \bar{p}) \to x$$

$$\Downarrow$$

$$1$$

Figure 7.4. An example of extraction and simplification.

Figure 7.4 illustrates the flow of interleaving extraction and simplification that reduces the constraint 7.18 into two det-constraints, with help from a third det-constraint extracted elsewhere. $\qquad \square$

7.5 The Overall Algorithm

The extraction and simplification algorithm is given in Figure 7.5. The main function (*extract*) takes a set of constraints, C, and returns the extracted det-constraints in E and the simplified constraints in C. The syntactical extraction (*synt_extract*) is applied first because it is fast and because it simplifies the constraints prior to BDD building. The ensuing functional extraction is iterative, as shown in the *do* loop. The set *currE* stores the det-constraints extracted in the current iteration, and *prevE* is the *currE* from the last iteration. In each iteration, each constraint f in C is simplified with respect to det-constraints in *prevE* using conditional substitution (*cond_subst*), and the result f' replaces f in C; then, for each input variable in f' that does not have an extraction, functional extraction using Theorem 7.7 is applied (*func_extract*). If there is an extraction, it is added to *currE*. At the end of the iteration, all extractions in *currE* are added to E, and all constraints in C are simplified using the BDD Restrict (*restrict*) operation, as discussed at the end of Section 7.3. The *do* loop terminates if *currE* is empty — this will happen within a finite number of iterations, since all constraints have a finite number of input variables.

7.6 Historical Perspective

The idea of extracting det-constraints stemmed from our observation of real-life design constraints in which inputs are very commonly assigned values stored in memory elements. A syntactical extraction was the natural choice at the conception of this idea. This attempt at a functional extraction was

```
Extract(C) {
    E = currE = ∅;
    (C, E) =Synt_extract(C);
    do
        prevE = currE;
        currE = ∅;
        foreach f in C
            f' =Cond_subst(f, prevE);
            C = C − {f};
            C = C ∪ {f'};
            foreach not-yet-extracted input variable x in f'
                eₓ =Func_extract(f', x);
                if (eₓ ≠ nil) currE = currE ∪ {eₓ};
            E = E ∪ currE;
            C =Restrict(C, currE);
    while (currE ≠ ∅);
}
```

Figure 7.5.   Det-constraint extraction.

inspired by the *state assignment extraction* work of Yang, Simmons, Bryant, and O'Hallaron [YSBO99]. The key result of their work is as follows:

THEOREM 7.10 (**[YSBO99]**) Let f be a Boolean formula, then

$$f \leftrightarrow (x \in g) \cdot h \qquad (7.26)$$

where x is a variable whose possible values are in the set L, and $h = \exists_x f$ and $g = \psi(t, h)$; ψ is a simplification function which uses the care set h to minimize t. The relation $t \subseteq h^{on} \times L$ is computed as

$$\bigvee_{l \in L} \left(ITE(f|_{v \leftarrow l}, \{l\}, \emptyset) \right). \qquad (7.27)$$

If t is also a partial function, i.e., each minterm in h^{on} corresponds to a unique value in L, then g is a function, and

$$f \leftrightarrow (x = g) \cdot h. \qquad (7.28)$$

However, there are no distinctions between state and input variables in their work, and the assignments are unconditional. We attempted to modify the above approach to meet our needs in the "natural" way, as given by the following theorem.

THEOREM 7.11 Let f be a Boolean formula, $k = \overline{\forall_x f}$ and $h = \exists_x f$. Let ψ be a simplification function which uses the care set h to minimize t. Let $e = f \cdot k$ and $g = \psi(e_x, \exists_x e)$. Then

$$f \leftrightarrow (k \rightarrow (x = g)) \cdot h. \qquad (7.29)$$

We needed to make sure that k and g did not depend on any input variables by applying care-set optimization. Even so, we failed to obtain some obviously extractable det-constraints. For example, $a + x$ in the constraint

$$f = (a + x) \cdot (b \cdot y + x) \qquad (7.30)$$

where a, b are the state variables, and x, y the input variables.

It turns out that the above method works only if f has a *conjunctive bi-decomposition* such that the intended input variable and the rest of the input variables belong to different conjuncts.[6] This is an obvious limitation.

Bertacco and Damiani [BD97] proposed a method to build the decomposition tree for a Boolean function from its BDD representation. Their method has a similar restriction that the variable supports of the components be disjoint, and, therefore, is not suitable for our application.

An earlier work by McMillan [McM96] gives similar results to those of Yang *et al.* His method uses the BDD *constrain* operator and can factor out *dependent variables* from Boolean functions. However, because the dependency is unconditional (i.e., in our case, for all state valuations), the method cannot be adopted for a complete extraction, either. The example in (7.30) above also showcases the inability of this method to extract all possible det-constraints.

It can be proven that a det-constraint on x is extractable from f iff there exists a conjunctive bi-decomposition of f such that one conjunct depends on x and some state variables, and the other can depend on all the variables. Our test in Theorem 7.7 detects exactly such a decomposition.

The most comparable works on similar decompositions are those of Bochmann *et al.* [BDS91], and Mishchenko *et al.* [MSP01], from the logic synthesis and optimization community. They provided the criteria to test various *groupabilities* of the variables in a function. For example, the variables in a function are conjunctive groupable if the function can be decomposed into the conjunction of two subfunctions, which can have common variables. This seemed to match our need of testing if there is a decomposition that separates the input being considered for extraction from the other inputs. However, test results vary with the choice of common variables. In our case, this can be any subset of the state

---

[6]To see this, let f be a constraint with the decomposition $(k \rightarrow (x = g)) \cdot h$ (7.29). Also, let there be some other input variables V in f. Therefore, V must be in h. In the meantime, x is not in h since $h = \exists_x f$ (Theorem 7.11). It follows that, in the above decomposition, the first conjunct contains exactly one input variable x, and the second conjunct h contains all the other input variables.

variables. Therefore, it can take multiple (in the worst case, exponential to the number of state variables) groupability tests to decide if a det-constraint exists for an input variable. In contrast, our approach needs just one test.

7.7   Experiments

7.7.1   Impact on Building Conjunction BDDs

The experiments are intended to compare the effects of det-constraint extraction on building BDDs for the partitioned constraints. Twelve commercial designs are used in the experiments. These designs include an ALU (Arithmetic Logic Unit) transaction level model, a crossbar switch, bus interfaces, and controllers for a program counter, pipeline, cache, and memory. These four configurations are compared:

no-extraction:	with no extraction
syntactical:	with the syntactical extraction
functional1:	with the non-recursive functional extraction
functional2:	with the recursive functional extraction

Table 7.1 demonstrates the effect of the three types of extractions. For this experiment only, the functional extractions are run without first applying syntactical extraction in order to perform the sanity check in which the former always subsumes the latter. Columns 1 and 2 give the numbers of constraints and input variables of the designs, respectively. The $\#e\_c$ and $\#e\_i$ columns give the numbers of constraints and input variables with extractions, respectively. The last two rows sum up the numbers in each column, and give the overall percentages of constraints and inputs that yield det-constraints. As can be seen, although det-constraints are rare in some designs (especially the ones with very few constraints), on average, the extraction methods still achieved extraction on 36% to 53% of constraints and 62% to 71% of input variables. It is obvious that the power of extraction grows noticeably in the order of syntactical, non-recursive, and recursive extractions.

Tables 7.2 through 7.5 compare the results of building BDDs for the partitioned constraints. The reported times and BDD node counts include the times and number of BDD nodes used in extraction. Dynamic variable reordering is enabled in all examples except in *rio*, where a fixed order is used in order to avoid BDD blowup. Table 7.2 shows the results of building conjunction BDDs without any extraction. Column 1 is the number of BDD conjunction operations performed during partitioning. Column 2 shows the number of resulting parts. Columns 3, 4, and 5 show the peak number of BDD nodes, the number of BDD nodes in the result, and the time for building the BDDs, respectively.

The impact of extractions on BDD building is shown in Tables 7.3, 7.4, and 7.5, respectively for *syntactical*, *functional1* and *functional2*. It should be clearly stated that the sharp increase in the number of parts between the

Table 7.1. Result of syntactical and functional det-constraint extractions.

circuit	circuit stats		syntactical		functional1		functional2	
	#cons	#input	#e_c	#e_i	#e_c	#e_i	#e_c	#e_i
mmq	117	207	18	25	42	49	77	53
qbc	93	174	58	169	75	174	75	174
qpag	215	283	149	282	173	282	197	283
qpcu	109	34	75	34	93	34	93	34
rio	198	371	80	283	89	289	96	292
sbs	108	423	95	422	96	423	97	423
aips	75	79	0	0	4	3	5	3
yapp	20	18	1	1	3	1	3	1
smi	74	233	47	223	59	224	60	224
sdca	135	673	83	600	98	606	124	664
lef	28	101	0	0	0	0	0	0
max	506	693	0	0	34	32	65	176
Total	1678	3289	606	2039	766	2117	892	2327
%			36%	62%	46%	64%	53%	71%

Table 7.2. Building constraint BDDs with no det-constraint extraction.

circuit	#conj	#part	peak	result	time
mmq	92	26	82782	24535	17.0
qbc	60	34	10220	2689	2.5
qpag	187	29	968856	142943	272.4
qpcu	94	11	14308	4563	1.0
rio	133	66	1299984	375723	3.3
sbs	69	40	12264	2940	1.5
aips	53	2	61320	31239	21.1
yapp	17	4	5110	818	0.1
smi	49	25	31682	9666	4.3
sdca	104	32	115486	58370	180.5
lef	23	6	44968	7310	0.1
max	432	69	441504	26336	90.5

functional and syntactical extractions is partially due to the fact that the former extracts signals bit by bit, while the latter can extract bus signals at once. Therefore, the effect of partition refinement is more realistically represented in the number of conjunctions performed among the simplified versions of the original constraints. Note that we do not conjoin the det-constraints. It can be seen that, as the extraction gets more powerful, the number of parts increases. Although the number of conjunctions decreases quickly in *syntactical* and more in *functional1*, we observed that *functional2* does not improve the number further, even though it has the most extractions. As a result, BDD sizes in *functional2* increase slightly over those in *functional1*. Obviously, in our examples, simplification from the extra extractions in *functional2* did not remove more input variables to refine the partition, although in theory it could. Overall, functional extractions show large improvements over their syntactical counterpart. In all examples, extracting det-constraints has a clear advantage in time and space usages over not extracting det-constraints.

Table 7.3. Building constraint BDDs with syntactical det-constraint extraction.

circuit	#conj	#part	peak	result	time
mmq	80	37	32704	9501	8.9
qbc	19	77	9198	1754	0.1
qpag	65	156	155344	89195	61.4
qpcu	24	84	4088	983	0.0
rio	97	127	1040396	149152	1.5
sbs	10	104	15330	1663	0.1
aips	53	2	61320	31239	21.5
yapp	16	5	5110	761	0.1
smi	23	53	16352	4710	2.0
sdca	49	89	49056	19931	31.5
lef	23	6	44968	7310	0.1
max	432	69	473186	26336	93.1

7.7.2   Impact on Simulation

Simulation directly benefits from finer partitions that usually result in smaller conjunction BDDs. Furthermore, det-constraints by themselves also contribute to the speedup of vector generation since they can produce quick solutions to input variables. Table 7.6 shows the speedup of random simulation due to det-constraint extraction. Non-recursive extraction was used in this experiment. Each design is simulated three times, with runs of 1000 cycles, using randomly generated inputs from the conjunction BDDs. Columns 1 and 2 report the average times spent in simulation generation from BDDs with and without extraction, respectively. Column 3 gives the ratio of generation time without extraction to that with extraction. The speedup is proportional to the reduction in BDD size when the conjunction BDDs are complex enough, for example, in *rio*, *qpag*, and *mmq*. For smaller BDDs, the overhead of handling finer partitions may offset the size reduction which, in the case of *qpcu*, actually

Table 7.4. Building constraint BDDs with non-recursive det-constraint extraction (*functional1*).

circuit	#conj	#part	peak	result	time
mmq	72	58	33726	9529	6.2
qbc	10	92	5110	1294	0.0
qpag	60	216	47012	20613	12.3
qpcu	17	136	4088	801	0.0
rio	94	140	1072460	142997	1.5
sbs	10	503	34748	1612	0.1
aips	52	8	53144	26261	27.1
yapp	14	7	5110	360	0.1
smi	23	66	11242	3016	1.9
sdca	48	107	45990	11220	41.8
lef	23	6	44968	7310	0.1
max	423	111	403690	23669	172.3

Table 7.5. Building constraint BDDs with recursive det-constraint extraction (*functional2*).

circuit	#conj	#part	peak	result	time
mmq	72	58	33726	9529	6.2
qbc	10	113	6132	1428	0.0
qpag	60	268	79716	28221	26.7
qpcu	17	146	13286	1042	0.0
rio	94	144	1226400	143071	1.6
sbs	10	528	49056	1812	0.1
aips	52	9	61320	20765	34.8
yapp	14	7	5110	360	0.1
smi	23	255	53144	2964	4.2
sdca	48	111	52122	11877	43.6
lef	23	6	44968	7310	0.1
max	423	263	449680	21206	178.0

increased the generation time. Nonetheless, our concern is more with the long generation times from large conjunction BDDs, in which cases we achieved a speedup ratio of about 2.5.

7.8 Summary

We have presented a method for simplifying constraint solving and demonstrated its effectiveness in the application of random simulation-vector generation. The source of the simplification is the refining of constraint partition by

Table 7.6.   Speedup of random simulation vector generation due to det-constraint extraction.

circuit	without extract	with extraction	speedup
mmq	6.98	4.57	1.53
qbc	1.55	1.24	1.25
qpag	6.46	2.57	2.51
qpcu	0.49	0.55	0.89
rio	302.63	132.00	2.30
sbs	2.12	1.98	1.07

extracting deterministic assignments to input variables. The result is a faster construction and a smaller size of the BDD representation of the constraints. The simulation-vector generation time is also reduced, both due to the smaller BDD size and the fast solution of input variables in det-constraints.

Chapter 8

MORE OPTIMIZATIONS

We have discussed test generation and optimization algorithms in previous chapters. In this chapter we describe additional optimizations and features in a constraint-based verification framework. The input to the framework consists of the design files, constraints, and randomization control, including input biases and distributions. The constraints are first compiled into a database. The generator engine operates on this database to provide inputs to the simulator. At each simulation cycle, the generator engine first samples the current state and determines the dynamic biases, then uses the state and bias information and the constraint database to generate a vector that is allowed by the constraints under the sampled state. The flow of the framework is shown in Figure 8.1.

The compilation stage is divided into several steps. Constraints are first levelized according to their priority as assigned by the user. A simplification is then applied to extract special constraints that can be solved independently

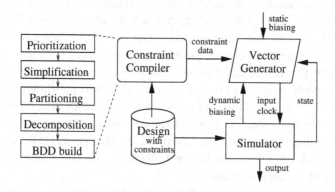

Figure 8.1.   A test generation framework.

and prior to the original constraints. These derived constraints are also used to simplify the original constraints. This simplification may help refine the constraint partitioning that occurs next – a process that breaks the constraints into groups. The constraints inside each group are then subjected to structural and functional decomposition for further simplification.

The vector generation engine has multiple choices of algorithms. The BDD-based *Simgen* algorithm given in Chapter 6 is fast and can generate vectors according to pre-determined distributions. The constraint synthesis approach can be both BDD- and SAT-based. It is faster than *Simgen*, provides a seamless synergy between simulation and formal verification, and is the choice for an emulatable testbench. However, randomization is not accurately controllable. *Simgen* and constraint synthesis are so-called off-line methods, because constraint solving is actually performed before simulation. All test generation needs to do, in these cases, is to retrieve the solutions, which takes an insignificant amount of time compared to the optimization and solving time. This makes *Simgen* and constraint synthesis desirable choices for fast simulation throughput. However, it is still possible for the quantity and complexity of constraints to grow beyond the reach of the BDD- and synthesis-based methods, making the solution database too large to be practical. Therefore, SAT-based, ATPG-based, or general purpose constraint solving methods, which trade time for space, should also be included in a robust constrained verification framework. We will discuss this topic in greater detail in Chapter 11.

In this chapter, we describe several extensions and optimizations to test generation used in the framework. Note that the optimizations are general to constraint solving problems and not specific to a particular type of generation engine.

8.0.1   Constraint Prioritization

Constraint prioritization stems from a frequently encountered situation where some constraints assume higher priority over others, because the inputs in the former can be decided solely using the first set of constraints. For example, in a bus protocol, one can choose a transaction type based on constraints that *define* the transaction type, then choose the attributes associated with that type based on constraints that define the attributes *using* the transaction type. This breaks up the constraints even before the disjoint-input-support partitioning is applied. In general, however, only the user knows which variables are constrained and which are used as parameters in a constraint. There are two ways to pass this knowledge to the constraint solver. The user can either prioritize the constraints, or specify which variables are to be constrained on what constraints. By default, all input variables are constrained.

A constraint prioritization is a partition of a set of constraints, C, with ordered components $\{C_1, \ldots, C_n\}$, where C_1 has the highest priority. This partition in turn defines a partition $\{X_1, \ldots, X_n\}$ of input variables X as follows. Let $sup(C_i)$ be the input support of C_i, then $X_1 = sup(C_1)$, and $X_i = sup(C_i) - \cap_{j=1}^{i-1} X_j$ for $1 < i \leq n$. X_i defines the set of input variables to be solved at priority level i, the set of input variables in C_i that have not been solved in a higher level.

EXAMPLE 8.1 Consider the constraints $x_1 + x_2$ of priority 1, $x_1 + x_2 + x3$ of priority 2, and $x_3 + x_4$ of priority 3. The inferred constrained variables for the three priority levels are $\{x_1, x_2\}$, $\{x_3\}$, and $\{x_4\}$, respectively. □

The second approach, by specifying the constrained variables, reverses the steps in constraint prioritization. For each constraint c_j, it marks a subset of input variables to be constrained, say x_j. Then all constraints that share constrained variables are grouped. This results in a partition of constraints, $\{C_1, \ldots, C_n\}$, and two sets of variable sets, $\{X_1, \ldots, X_n\}$ and $\{X'_1, \ldots, X'_n\}$, where X_i's and X'_i's are the constrained and parameter input variables in C_i, respectively. Unlike the first approach, which gives a linear order of C_i's, this approach gives a set of partial orders: C_i is solved before C_j if $X_i \cap X'_j \neq \emptyset$. Note that this can potentially lead to cyclic ordering, which we flag as being illegal. The acyclic partial orders are then collapsed into a linear order.

EXAMPLE 8.2 Consider the constraints $x_1 + x_2$, $x_1 + x3$, and $x_2 + x_4$. They have constrained variables $\{x_1, x_2\}$, $\{x_3\}$, and $\{x_4\}$, respectively. The inferred partial orders are: $x_1 + x_2$ before $x_1 + x3$, and $x_1 + x_2$ before $x_2 + x_4$. □

Now, with ordered partitions $\{C_1, \ldots, C_n\}$ and $\{X_1, \ldots, X_n\}$, test generation is first applied to C_1, then to C_2 conditioned upon the assignment to X_1, then to C_3 upon the assignments to X_1 and X_2, and so on. Obviously, the distribution of vectors so generated is determined by the product of the constrained probabilities of the assignments to $X_1, \ldots, X_n$.

Constraint prioritization is not only necessary for enhanced expressiveness of constraints, but also helps to divide complicated constraint systems into manageable blocks, which is beneficial to both constraint solving and debugging. We have observed the widespread use of constraint prioritization among *Simgen* users; many sampled designs involves more than three levels of constraints, some involve up to ten levels.

8.0.2 Tree-decomposition

Tree-decomposition (TD) [RS86] is a simplification technique extensively studied in Constraint Satisfaction Problems (CSPs) (e.g., [GLS99]) and in the similar conjunctive query problem in relational databases (e.g., [GJC94]). We show how to apply TD to refine the constraint partitions obtained with prioritization and disjoint-input-support.

The definition of TD varies with different applications. The following definition is aimed at minimizing the number of variables occurring in a BDD [YPAA03].

Let $C = \{f_1, \ldots, f_n\}$ be the set of constraints, X the set of input variables in C, and $var(f_i) \subseteq X$ the set of input variables in f_i. A TD of C is a tree $T = (I, E)$, with a set $var(i) \in X$ labeling each vertex $i \in I$, such that

1. For each constraint $f_i \in C$, there is a vertex $i \in I$ such that $var(f_i) \subseteq var(i)$

2. For each variable $v \in X$, the set of vertices $\{i \in I \mid v \in var(i)\}$ induces a (connected) tree

The width of a TD, usually referred to as the *tree-width*, is the maximum size of $var(i)$ over all $i \in I$. Finding the TD with the smallest width among all TDs of a constraint system is expensive so we use the linear time *triangulation* algorithm that heuristically searches for a TD with a small width [TY84]. Correctness of tree-decompositions in general is founded on the *jointness* of variables, that is, the projection of the tree to each variable gives one joint sub-tree. Some tree-decompositions use repetitive constraints to satisfy this property. Our approach instead uses free variables to play this role so that each constraint appears exactly once in the tree.

Before the tree can be used for constraint solving, it must be made *arc-consistent* [BFMY83]. Arc-consistency between a pair of adjacent nodes requires that any solution of the constraints from one node must have a compatible solution from the other, if the other node has a solution at all. In our case, arc-consistency is obtained, in one direction, as follows. First, we choose a node with a minimum variable support as the root of the tree. Then, starting from each of the leaf nodes, for each pair u and v, where u is a parent of v, we conjoin the constraints in u with the projection of the constraints in v onto the overlapping variables between u and v. Then the process is repeated for the parent u and each of its parents, and so on.

In constraint solving, nodes are solved one at a time in the top-down order, starting from the root. The solution at each node is applied to that of the next. Note that the number of variables involved in arc-consistency computation and constraint solving is upper bounded by the tree-width.

EXAMPLE 8.3   Figure 8.2 demonstrates a tree-decomposition of nine constraints and the joint subtrees of the supporting variables (subtrees of one node, e.g, of variable node g, are not shown). Each constraint is represented with its set of constrained variables. The width of the tree is three, determined by the nodes $\{a, b, d\}$ and $\{b, d, f\}$. Without tree-decomposition, these nine constraints would have to be collapsed into one conjunctive constraint of nine variables and solved together. This 1:3 (tree-width to total) reduction ratio is not uncommon in real world examples, as shown in Table 8.1 in the experiment.   □

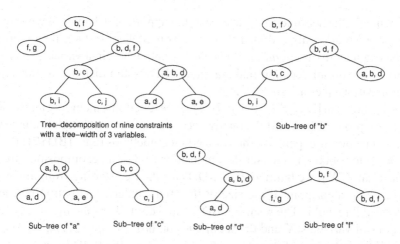

Figure 8.2. An example of tree-decomposition.

The effect of TD can be seen in Table 8.1, where the "time" and "size" columns are for the BDD build time (in seconds) and final BDD size (in 1000 nodes), respectively. The last column gives the ratio of the tree width to the total number of variables, indicating the reduction of variable count in BDDs after TD.

Table 8.1. Effect of tree-decomposition.

design	without TD		with TD		
design	time	size	time	size	ratio
qpag	103.1	200.3	0.09	2.65	0.25
sqbc	0.01	6.13	0.00	2.08	0.40
qpcu	0.02	3.07	0.00	0.79	0.65
mmq	14.16	6.60	0.16	1.33	0.30
rio	0.97	142.7	1.53	133.9	0.66
ccu	t/o	-	74.1	4486.2	0.50

8.0.3 Functional Decomposition

The *Simgen* framework provides several previously described methods for simplifying constraint solving. In extreme cases, we need even heavier preprocessing to make vector generation sufficiently fast for simulation. One possibility is to try to break the large constraints using *disjunctive functional decomposi-*

tion, wherein the subfunctions have disjoint support. In Chapter 6, we showed a vector generation algorithm that operates on BDDs. In fact, the algorithm can be extended to any *read-once* decision graph, wherein a variable is visited at most once on all decision making paths. BDDs and disjunctive functional decomposition trees are such examples.

It was shown in [DMvS97] that *fully sensitive* functions, including all Boolean functions, provide the finest disjunctive decomposition. Although finding such a decomposition is expensive, there are efficient methods (e.g., [BDS91, BD97, BMD02]) for obtaining coarser decompositions.[1] For a decomposition tree of a constraint $f(X)$, we build a set of BDDs for the tree starting from the leaves and inserting a cut-point if the size of the BDD under construction reaches a predefined threshold. The resulting BDDs, in the set $G = \{g_0, g_1, \ldots, g_n\}$, are functions of variables X and the set of cut-points $Y = \{y_1, \ldots, y_n\}$, where g_0 is the BDD at the root of the tree, and g_i $(1 \leq i \leq n)$ the BDD for cut-point y_i.

To apply the previous vector generation algorithm to this decomposition tree, the weight computation in the equation (6.2) in Chapter 6 needs to be augmented by adding, to the definition of the probability p, the case "if x is a cut-point corresponding to BDD g_x, then p returns the weight of the root of g_x." Also, the random walk used to generate vectors needs to be augmented as follows. If the node being visited is associated with a cut-point c_i, then, after assigning c_i according to the branching probability, the traversal recurs into the corresponding BDD g_i before it explores the selected branch in the current BDD. The random walk in g_i is done similarly but with an important difference: if c_i was assigned to 0, then each node in g_i will have its branching probability replaced by its complement (of 1).

8.0.4   Formula Factorization

Formula factorization, in our context, is the process of deriving a conjunctive decomposition with disjoint input support. It is a special case of the more general functional decomposition. Consider the factorization of a constraint formula f. Ideally, we want to find the decomposition $f = f_1 \wedge f_2 \wedge \ldots \wedge f_n$ where the (constrained) input variable supports in the factors f_i constitute the finest partition of input variables. This is generally a covering problem. A cover of a Boolean formula f is a set of implicants of f. A prime implicant of f is one that implies no other implicants of f. An *Irredundant Prime Cover* (IPC) is a prime cover that has no proper subset that is also a cover. Factorization is the dual problem of finding a cover.

In [Hav01], the *Blake canonical form* [Bla38] of propositional formulas is used as the starting point for factorization. The Blake form of a propositional

---

[1] A more conservative approach is to transform the circuitry behind the constraint into a tree by collapsing the *fanout-reconvergent regions*.

formula is the set of all prime covers (in clauses) of the formula. The Blake form can be computed, for example, by "factoring out" any sum-of-products form for the formula. To derive a minimal factorization, redundant clauses from the prime cover must be removed in a manner such that only the clauses with the smallest input-variable supports are kept. The process repeatedly applies resolution among pairs of clauses, removing clauses subsumed by the resolvents. It selects clauses for resolution in the order of the increasing number of input literals and resolves only on the state variables. The process terminates when no more clauses can be removed. To obtain the final factorization, the remaining clauses are grouped according to overlapping input-variable support.

However, Blake canonical forms for complex constraints are not very manipulatable. In these cases, a potentially coarser factorization can be obtained using the *meta-product* approach [CM92, CMFT93]. This approach is BDD-based and computes an IPC implicitly, which avoids both the canonical form construction and the iterative resolutions [YH01].

8.1 Implication of Multiple Clocks

System components often operate under their local clock. *Simgen* handles multiple clock domains by introducing clock constraints for clock generation and clock qualifiers that govern normal (input) constraints. Clock constraints often involves clock generation circuitry, e.g., state machines, but they differ from the traditional explicit generation method by being able to introduce randomness in the inter-relationship of clocks. This is an important requirement for verification in multiple clock domains. Clock-qualified constraints are the natural consequence of a multiple-clock design, since each constraint should only be sampled by the clock governing the involved signals. The new constraints are handled using the same core procedure, but the sequence of sampling state and solving constraints will be different, as shown in Figure 8.3.

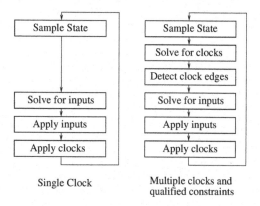

Figure 8.3. Events in each simulation cycle.

8.2 Summary

We have discussed several optimization techniques and the adoption of multiple clocks in test generation. Constraint prioritization is based on the natural hierarchies existing in a system of constraints, whereas tree-decomposition infers a hierarchical solving order from the structure of the constraints. While these two optimizations aim at breaking constraints into smaller groups based on user input and structure information, decomposition and factorization attempt to achieve the same effect through functional analysis. These optimization techniques, when combined with disjoint-input-support partition (Chapter 6) and constraint extraction and simplification (Chapter 7), provide powerful tools for performance enhancement in a constraint-based test generation framework.

Chapter 9

CONSTRAINT SYNTHESIS

Constraint synthesis is an alternative to the constraint solving method discussed in Chapter 6. Instead of retrieving solutions from BDDs, constraint synthesis derives, from a given constraint, a set of functions with a composed range identical to the solution space of the constraint. In other words, if those functions are used to drive the design inputs, the behavior of these inputs is exactly what is specified by the constraint.

There are multiple reasons why we are interested in constraint synthesis. For one, it enables a seamless integration of a hybrid simulation and formal verification flow, as proposed in Chapter 1. As part of the effort devoted to such a flow, we have shown in Chapter 4 that temporal constraints and assertions can be used interchangeably by first breaking them down into FSMs and Boolean formulas. Such FSMs can function as generators or monitors, depending on the usage of those Boolean formulas. Monitoring these formulas is straightforward and essentially the same in both simulation and formal verification. However, input generation still takes fundamentally different approaches in the two types of verification, that is, the formulas solved as constraints in simulation versus assumptions used directly in formal methods. The former requires the simulator to either work with a *Simgen*-like tool with an interface overhead, or has a built-in constraint solving capability. This dissimilarity in treating constraints is eradicated by constraint synthesis. The synthesized functions, directly translatable to circuits, are handled natively by any commercial HDL simulators and formal tools.

While constraint synthesis is a desirable for the sake of the effortless integration of simulation and formal verification, it is far more useful in emulation. In previous times, stimulus was applied to emulated hardware in two ways: through a software testbench sitting in a computer, or as pre-generated vectors loaded at the beginning of emulation. Both have drawbacks. The communication overhead between the emulation hardware and computer at each simulation cycle would defeat the very purpose of emulation, being a faster simulator, whereas, in the pre-generation approach, the intractability between the design and the testbench is lost. The missing piece in this puzzle is the so-called *synthesizable testbench* that executes in hardware together with the design. For constraint-based testbenches, constraint synthesis fits the bill nicely.

What we have not addressed is the efficiency of test generation with constraint synthesis. It turns out that, in theory, constraint synthesis-based generation is always faster than constraint solving with BDDs, and uses about the same amount of memory. Of course, the speed difference between the two will be much more significant in emulation, because the former is restricted to running on a computer and each solution takes multiple instructions, whereas the latter is done in one clock cycle in emulation.

The drawback of constraint synthesis is that controlled randomization becomes more difficult than constraint solving in BDD. This is attributed to the very nature of constraint synthesis — the mapping of non-solutions to solutions.

More specifically, constraint synthesis deals with the following problem: for a constraint $f : B^n \mapsto B$, derive a mapping $B^n \mapsto B^n$ whose range is the onset of f. This formulation is equivalent to the *Boolean Unification* (BU) problem [W. 87, MN89] and falls in the general category of parametric Boolean equation solving (e.g., [CM90, FTKC91, MDA96]). In this chapter, we discuss a constraint synthesis method based on BU, called *Consynth* [YAAP03]. Randomization methods are discussed afterward. Like the test generation algorithms in *Simgen* 6, *Consynth* also manipulates constraints and logic operations using BDDs. *Consynth* extends BU in two aspects. First, it simplifies the solution by utilizing don't care information present in hardware constraints involving both input and state variables, and in multi-level logic. Second, further optimization is achieved by heuristically removing parametric variables.

We will also cover other related methods, including Coudert and Madre's method based on the *Constrain* operation [CM90, CBM89], Kukula and Shiple's method of constructing circuits from relations [KS00], and a sketch of an SAT-based implementation of *Consynth*. At the end of this chapter, we give experimental results and comparisons.

9.1   Problem Formulation

Let $f : \mathbb{B}^{m+n} \mapsto \mathbb{B}$ of input variables $X = \{x_1, \ldots, x_n\}$ and state variables $Y = \{y_1, \ldots, y_m\}$ be the constraint of concern. Denote the set of *ground solutions* to $f(X, Y) = 1$ by S, where

$$S = \{\{\alpha, \beta\} \mid \alpha \in \mathbb{B}^n, \ \beta \in \mathbb{B}^m, \ f(\alpha, \beta) = 1\}. \tag{9.1}$$

Let

$$S_\beta = \{\alpha \in \mathbb{B}^n \mid \exists \beta(\{\alpha, \beta\} \in S)\} \tag{9.2}$$

be the set of input vectors pertaining to state β in S. A state β is *legal* in f iff $S_\beta \neq \emptyset$. Let

$$L = \{\beta \in \mathbb{B}^m \mid S_\beta \neq \emptyset\} \tag{9.3}$$

be the set of legal states in f. We adopt Rudeanu's [Rud74] terminology, with modifications to accommodate states, for describing different kinds of solutions to equations: A vector of functions $F : \mathbb{B}^{n+m} \mapsto \mathbb{B}^n$ is called a *solution* to $f(X, Y) = 1$ iff

$$\forall \beta \in L, \; F(\mathbb{B}^n \times \{\beta\}) \subseteq S_\beta, \tag{9.4}$$

F is called a *general solution* iff

$$\forall \beta \in L, F(\mathbb{B}^n \times \{\beta\}) = S_\beta, \tag{9.5}$$

and F is called a *reproductive solution* iff F is a general solution, and

$$\forall(\beta \in L, \alpha \in S_\beta), \; F(\alpha, \beta) = \alpha. \tag{9.6}$$

More intuitively, we apply a random but **legal** state vector and a random input vector to a set of functions F, which produces an input vector. We then apply this input vector, plus the random state vector, to the constraint f to see if it is satisfied. F's relation to f is then defined with respect to the outcome of this test. F is a solution of f if F can produce input vectors satisfying f and no other input vectors; F is a general solution of f if F can produce all input vectors satisfying f and no other input vectors; a reproductive solution is a special general solution such that F maps any input vector satisfying f to that input vector itself. Note that for illegal states the solutions are left undefined.

THEOREM 9.1 Let F be a reproductive solution to the equation $f(X, Y) = 1$, and L the set of legal states of $f(X, Y)$. Then, for any solution G of this equation,

$$G(\mathbb{B}^n \times L) = F(G(\mathbb{B}^n \times L)).$$

Both general and reproductive solutions are complete and sound in representing the full solution space of constraints. The reproductive solution gives us a higher level of certainty about how a random input is mapped to a legal input: we know at least that legal inputs map back to themselves.

EXAMPLE 9.1 Consider the constraint, "if a state condition y_1 holds, then inputs x_1, x_2, and x_3 are one-hot encoded":

$$y_1 \; \to \; x_1 x_2 \bar{x}_3 + x_1 \bar{x}_2 x_3 + \bar{x}_1 x_2 x_3.$$

A reproductive solution of this constraint is

$$x_1 = \bar{y}_1 p_1 + (y_1 + p_1) \bar{p}_2 \bar{p}_3$$
$$x_2 = (\bar{y}_1 + \bar{p}_3) p_2$$
$$x_3 = p_3$$

where p_1, p_2 and p_3 are the free parametric variables. We will show how to compute it later. The valuation of state variable y_1 partitions the solution space into two regions, one for $y_1 = 1$ where the one-hot encoding of the inputs is asserted, and one for $y_1 = 0$ where all combinations are possible. There is no illegal state.

We verify that the above is indeed a reproductive solution by confirming that

1. all and only the legal input vectors are generated, and

2. the solution functions map a legal input vector to itself.

In the case where $y_1 = 0$, the solution functions simplify to

$$x_1 = p_1, \quad x_2 = p_2, \quad x_3 = p_3.$$

That the above is a reproductive solution should be obvious. In the case $y_1 = 1$, the solution functions simplify to

$$x_1 = \bar{p}_2 \bar{p}_3, \quad x_2 = p_2 \bar{p}_3, \quad x_3 = p_3.$$

The mapping from parametric to input vectors is shown in Figure 9.1. As can

p1 p2 p3	x1 x2 x3
0 0 0	
0 0 1	0 0 1
0 1 0	0 1 0
0 1 1	
1 0 0	1 0 0
1 0 1	
1 1 0	
1 1 1	

Figure 9.1.   Example: mapping of a reproductive solution.

be seen, all and only the three one-hot vectors are on the input vector side, and each of those one-hot vectors maps to itself.   □

9.2   The Constraint Synthesis Method

We will describe a synthesis method, *Consynth*, that finds a reproductive solution for a given constraint. We will start with the core synthesis procedure, then address two types of optimizations: one through the use of don't care information, the other with the elimination of variables introduced in the derivation of the general solution.

9.2.1 Deriving Reproductive Solutions

Let f be a constraint of input variables $X = \{x_1, \ldots, x_n\}$ and state variables Y. Denote the *projection* of f onto $\{x_i, \ldots, x_n, Y\}$ by f^i, which is given by the existential quantification $\exists_{x_1, \ldots, x_{i-1}} f$, for $1 \le i \le n$. Note the special case $f^1 = f$. The notion of projection is critical to the synthesis procedure because of the following theorem.

THEOREM 9.2 Given a constraint f and a state, either all, or none, of the projections $f^1, \ldots, f^{n+1}$ are satisfiable.

PROOF: The theorem follows directly from the definition of projections, which are existential quantifications of f. □

Note that, by definition, projection f^1 is f. Hence, given a state valuation and a canonical representation of f, such as the one in BDD, we can quickly determine whether f is satisfiable (in time $O(n+m)$), and therefore whether all of its projections are satisfiable. In addition, if there is a satisfying assignment, say π, to variables in f^i, then there exists an assignment to x_{i-1}, which together with π satisfies f^{i-1}. This can be generalized to the following theorem.

THEOREM 9.3 Under a legal state, any input assignment satisfying f^i is a suffix of an input assignment satisfying f^j, for all $j < i$.

PROOF: See Appendix. □

The above theorem indicates that we can construct a solution for f by successively solving for inputs in the projections. We start from f^n, which contains only x_n and the state variables, then apply the solution of x_n to f^{n-1}, in which the only inputs are x_n and x_{n-1}. We then solve for x_{n-1}, and so on. Since $f^1 = f$, when we solve x_1, we also obtain a solution for f. So what is left to be decided is how we compute each input in a projection.

Note the following decomposition of a function g over an orthogonal basis $\{g_x g_{\overline{x}}, g_x \overline{g_{\overline{x}}}, \overline{g_x} g_{\overline{x}}\}$:

$$g = g_x g_{\overline{x}} + g_x \overline{g_{\overline{x}}} \, x + \overline{g_x} g_{\overline{x}} \overline{x}. \tag{9.7}$$

Suppose a solution exists for $\exists_x g$. Denote it by a vector α. Then, a satisfying assignment to x in g is chosen as

$$x = \begin{cases} u & \text{if } \alpha \in g_x g_{\overline{x}} \\ 1 & \text{if } \alpha \in g_x \overline{g_{\overline{x}}} \\ 0 & \text{if } \alpha \in \overline{g_x} g_{\overline{x}}. \end{cases} \tag{9.8}$$

This solution to x is *general* in that u, a free variable, has the option of assigning either 0 or 1 to x whenever it is allowed. However, an extra degree of freedom can be introduced if we consider the states under which no input combinations are allowed by the constraint, that is, the states specified by $\overline{g_x}\,\overline{g_{\overline{x}}}$. It should be stressed that, although for simplicity we define all such states as being illegal, some of them may actually never occur in a design, providing a don't care space for optimization. For example, most of the combinations of one-hot encoded state variables are unreachable by design; control and status registers may be confined to a smaller space than designed by the configuration being currently tested. To be utilized, these conditions have to be given as constraints. Constraints like those are not for input generation, but rather are used as assumptions. Recalled constraints depending only on state are checked as properties. Here we assume all illegal states are unreachable and use them for optimization. Meanwhile, illegal states will be flagged in test generation. We have deferred the details on don't care optimization and the detection of illegal states to the coming subsections.

Returning to solving inputs in the projections, we see that the above analysis applies by corresponding x_i, the variable to be solved in f_i, to x in g, and f^{i+1} to $\exists_x g$ since $f^{i+1} = \exists_{x_i} f^i$. Therefore, according to Equation (9.8) and the don't care condition, we come to the following general solution for x_i:

$$\sigma_i(x_i) := (f^i_{x_i} f^i_{\overline{x_i}}) u_i + (f^i_{x_i} \overline{f^i_{\overline{x_i}}}) + (\overline{f^i_{x_i} f^i_{\overline{x_i}}}) d_i. \tag{9.9}$$

Note that u_i is the free variable introduced for x_i, and that d_i can be any function that takes advantage of the don't care condition. Conceive of the σ_i's as logic circuits with inputs u_i's and outputs x_i's. A general solution to $f(x_1, \ldots, x_n, Y)$ can then be constructed as a multi-level and multi-output circuit that is the cascade of σ_i's according to the following rules:

1. x_1 is the output of σ_1
2. x_i is the output of σ_i, whose input u_j is connected to x_j for $1 \leq j < i$
3. all state inputs to σ_i for $1 \leq i < n$ are connected to the corresponding state lines from the design

Figure 9.2 gives an example of a 3-output solution circuit for a 3-input constraint, where $\sigma_1, \sigma_2, \sigma_3$ are the solutions, and u_1, u_2, u_3 and Y are the input and state variables, respectively. The following theorem summarizes the property of constraint synthesis we have discussed so far.

THEOREM 9.4 The solutions σ_i's connected as above defines a mapping $\sigma :$ $\mathbb{B}^n \mapsto \mathbb{B}^n$, such that for any legal state and any pair of input vectors $\alpha, \alpha' \in \mathbb{B}^n$ and $\sigma(\alpha) = \alpha'$

1. α' satisfies f, and

2. if α satisfies f, then $\alpha' = \alpha$.

PROOF: See Appendix. □

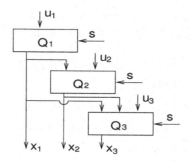

Figure 9.2. Example: Synthesis of a 3-input constraint.

9.2.2 Don't Cares

The don't care optimization arises from the introduction of the term

$$\overline{f^i_{\overline{x_i}} f^i_{x_i}} d_i \qquad (9.10)$$

in the solution for each input x_i in (9.9). The careset actually has two sources, the legal states and the limited input combinations observed by the general solutions σ_i's. For example, in Figure 9.2, the first two inputs of the circuit σ_3 are constrained by the outputs of σ_1 and σ_2.

We know that, for a constraint $f(X, Y)$ (recall that X and Y are the input and state variables, respectively), the legal state space is given by $\exists_X f$. In addition, under a legal state s, the only input patterns (excluding the free variable u_i) allowed for a solution σ_i are those satisfying the projection f^{i+1} under s, i.e., $\exists_{x_i} f^i|_s$. Overall, the careset over the input-state space is the product of the two, reducing to $\exists_{x_i} f^i$. The inversion of this, the don't care condition, equals the condition for d_i in (9.10). This is not an accident but a result of the synthesis.

To take advantage of this careset information and to realize d_i, we simplify the solution in (9.9) by first setting $d_i = 1$, which gives

$$\sigma_i(x_i) = f^i_{x_i} u_i + \overline{f^i_{\overline{x_i}}}. \qquad (9.11)$$

This reduction is intended to minimize intermediate BDD operations, which tend to explode. Then we optimize the result again with respect to the careset $\exists_{x_i} f^i$ using the BDD *restrict* operator [CM90].

EXAMPLE 9.2 We revisit the example given in Section 9.1 and show how the reproductive solution is obtained. The constraint is

$$y \;\rightarrow\; x_1 x_2 \overline{x}_3 + x_1 \overline{x}_2 x_3 + \overline{x}_1 x_2 x_3.$$

Let us solve in the order x_1, x_2, and x_3, for which we use the parametric variables p_1, p_2, and p_3, respectively. For x_1, we have

$$f^1_{x_1} = \bar{y}_1 + \bar{x}_2\bar{x}_3 \text{ and } f^1_{\bar{x}_1} = \bar{y}_1 + x_2\bar{x}_3 + \bar{x}_2x_3.$$

From Equation (9.11), we get

$$\sigma_1(x_1) = (\bar{y}_1 + \bar{x}_2\bar{x}_3)p_1 + \overline{\bar{y}_1 + x_2\bar{x}_3 + \bar{x}_2x_3}$$
$$= \bar{y}_1p_1 + (y_1 + p_1)\bar{x}_2\bar{x}_3 + y_1x_2x_3.$$

For x_2, we have

$$f^2 = f^1_{x_1} + f^1_{l x_1} = y_1 + \bar{x}_2\bar{x}_3 + x_2\bar{x}_3 + \bar{x}_2x_3, \text{ and}$$
$$f^2_{x_2} = \bar{y}_1 + \bar{x}_3 \text{ and } f^2_{\bar{x}_2} = 1.$$

So the solution for x_2 is

$$\sigma_2(x_2) = (\bar{y}_1 + \bar{x}_3)p_2.$$

Finally, for x_3, we have

$$f^3 = f^2_{x_2} + f^2_{l x_2} = 1, \text{ and}$$
$$f^3_{x_3} = f^3_{\bar{x}_3} = 1.$$

Therefore,

$$\sigma_3(x_3) = p_3.$$

However, the final solution should depend on state and parametric variables only. After substituting the input variables out in the σ_i's, we get the reproductive solution:

$$\sigma_1(x_1) = \bar{y}_1p_1 + (y_1 + p_1)\bar{p}_2\bar{p}_3$$
$$\sigma_2(x_2) = (\bar{y}_1 + \bar{p}_3)p_2$$
$$\sigma_3(x_3) = p_3.$$

Because of the backward recursive substitution, the real solve order is the reverse of the order in which we solved the σ_i's, that is, x_3, x_2, x_1. □

One may be tempted to *collapse* the solutions for simplification, as shown in the last step of the previous example. This is actually what happens in standard BUs, like the one in [MN89]. In a sense, substitution can also exploit the input restriction we exploited above. In general, however, substitution can affect the size of the result in either way. Moreover, when there are many input variables, recursive and accumulative substitution in BU can cause BDD blowup. This was observed in our experiments.

9.2.3 Variable Removal

The problematic recursive substitution has one good effect: it eliminates the variables being substituted. However, the same effect can be achieved simply by recycling the input x_i whenever a free variable u_i is needed. This is sound due to the cascade of individual solutions shown in Figure 9.2, in which each u_i can be safely replaced by x_i.

Furthermore, the x_i's do not always have to be reintroduced. It is well known that function f is independent of variable x iff the *Boolean difference* $f_x \overline{f_{\overline{x}}} + \overline{f_x} f_{\overline{x}}$ (denoted by $\partial f / \partial x$) is 0. Applying this to the solution in (9.9), we have the following condition for the removal of x_i:

$$(f_{x_i}^i f_{\overline{x_i}}^i = 0) \bigwedge (\overline{f_{\overline{x_i}}^i f_{x_i}^i}(\partial d_i / \partial u_i) = 0). \tag{9.12}$$

As we have chosen d_i as 1 before the careset optimization, this condition can be simplified to

$$f_{x_i}^i f_{\overline{x_i}}^i = 0. \tag{9.13}$$

If this condition is met, the solution in (9.9) reduces to

$$\sigma_i(x_i) := \overline{f_{\overline{x_i}}^i}, \tag{9.14}$$

which is subjected to further optimization using the careset $\exists_{x_i} f^i$.

EXAMPLE 9.3   Consider the constraint $y_1 x_1 + \overline{y}_1 y_2 \overline{x}_1$, where y_1 and y_2 are state variables and x_1 an input variable. Since $f_{x_1}^1 = y_1$ and $f_{\overline{x}_1}^1 = \overline{y}_1 y_2$, then $f_{x_1}^1 f_{\overline{x}_1}^1 = 0$. Therefore, the solution for x_1 is $\overline{f_{\overline{x}_1}^1} = y_1 + \overline{y}_2$, which contains no parametric variables.   □

The satisfiability of the condition in (9.13) is determined by the constraint and the order in which the variables are solved. Therefore, instead of following the arbitrary order $x_n, \ldots, x_1$, we use the following heuristic to select a variable to solve in the current constraint projection f^i. First, compute the *removable set* of variables that meet the condition in (9.13). If this set is not empty, choose from the set a variable x such that $\exists_x f^i$ has the largest removable set. Otherwise, choose the highest ranked (in BDD variable ordering) input in f^i. Note that since the removable sets shrink monotonically (because for any variable ordering, $f_{x_i}^i f_{\overline{x_i}}^i = 0$ only if $f_{x_i} f_{\overline{x_i}} = 0$ initially) and they are relatively small compared to the whole set of inputs, finding a good ordering incurs only an insignificant overhead.

9.2.4   The Overall Algorithm

The algorithm that combines synthesis and optimization is given in Figure 9.3. Initially, the argument for the removable set, *r_set*, is passed in as *nil*; *GetRset(f)* computes the removable set of f, and the general solution for input x is computed as σ_x.

Two minor issues are not addressed in the algorithm. The first is the detection of illegal states. This is done by evaluating the current state against the set of all legal states $\exists_X f$. The second is that some input variables may remain

```
// GetRset () returns a set of variables
// satisfying (9.13)

Consynth(f, r_set) {
    if (f contains no input variables)
        return;
    if (r_set == nil)
        r_set = GetRset(f);
    if (r_set is empty)
        let x be the input in f with
            the highest rank (in BDD var. ordering);
    else
        let x ∈ r_set be such that ∃ₓf has the
            largest removable set;
        r_set = GetRset(∃ₓf);
    σₓ = xfₓ + f_x̄;
    σₓ = BDD_Restrict(σₓ, ∃ₓf);

    Consynth(∃ₓf, r_set);
}
```

Figure 9.3.   Constraint synthesis.

unprocessed when the algorithm terminates upon f becoming a constant (containing no input variables). We treat all the unprocessed input variables as free variables. This is clearly valid if the algorithm terminates with $f = 1$. If $f = 0$, then all states are illegal, and this will be caught by the illegal state detection.

9.3   Other Synthesis Methods

Consynth is an extension to Boolean Unification (BU) based on Boole's method. BU is a method for finding solutions for equations of Boolean functions. A solution σ to equation $f = g$ is a substitution to variables in f and g, such that $\sigma \circ f \Leftrightarrow \sigma \circ g$. For this reason, σ is also called a *unifier* of f and g. Since the equation can be rewritten as $f \oplus g = 0$, it suffices for BU to consider only the form $f = 0$.

There are several approaches to BU. Büttner and Simonis [W. 87] presented a BU algorithm that computes the *most general unifier mgu* [Sie84], from which any other unifier can be derived as an instance. Note that the reproductive solution produced by *Consynth* is an *mgu*. Martin and Nipkow [MN89] provided an historical perspective on the algorithm and traced the origin back to Boole himself. They also presented a second BU algorithm attributed to Löwenheim [Löw08]. In general, this algorithm produces a solution more complex than that of Boole's method [MN89].

The distinction of two sets of variables (state and input) is not a concern in those references. This distinction is addressed by Fujita *et al.* [FTKC91] in their application of Boole's BU algorithm to several logic synthesis examples, including the minimization of Boolean relations. Chen in [Che93] applied a similar algorithm to *Boolean matching*, also known today as logic equivalence checking. In their applications it is necessary to find a solution; their algorithm aborts if a solution is not attainable. In our case, unsatisfiability because of illegal states is treated as a don't care and used to minimize the overall solution.

In the following sections, we discuss several constraint synthesis methods orthogonal to BU.

9.4 Coudert and Madre's Method

Coudert and Madre [CM90] showed that the image of a set can be computed as the range of the transition functions *Constrained* with respect to the set. In addition, the resulting functions form a reproductive solution to the characteristic function representing that set. This fact is used by Coudert and Madre to generate "functional vectors" in [CBM89]. The *Constrain* operation, which we will refer to as *Generalized Cofactoring* (GC), is defined below.

GC takes two functions f and g of variables $\{x_1, \ldots, x_n\}$, and returns a function which agrees with f in the onset of g. The operation is a mapping decided by g and the variable ordering. The latter determines the "distance" between two vectors, as defined in the following.

DEFINITION 9.5 Let $x_1 \prec \ldots \prec x_n$ be the variable ordering. Let $\alpha = \alpha_1 \ldots \alpha_n$ and $\beta = \beta_1 \ldots \beta_n$ be two minterms. The distance between α and β, in symbols $\|\alpha - \beta\|$, is given by:

$$\|\alpha - \beta\| = \Sigma_{i=1}^{n} |\alpha_i - \beta_i| \cdot 2^{n-i}. \tag{9.15}$$

This definition of distance reflects the dissimilarity between the two vectors quantified by the variable ordering. We now define GC precisely. The implementation of GC in BDD is given in Figure 9.4.

DEFINITION 9.6 Given functions f, g, and a variable ordering, GC of f with respect to g, in symbols $f \downarrow g$, is defined by:

$$(f \downarrow g)(\alpha) = \begin{cases} f(\alpha) & \text{if } g(\alpha) = 1 \\ f(\beta) & \text{if } g(\alpha) = 0 \end{cases} \tag{9.16}$$

where $g(\beta) = 1$ and $\|\alpha - \beta\|$ is minimum under the given variable ordering.

```
f ↓ c {
    assert(c ≠ 0));
    if (c = 1 || f = 1 || f = 0) return f;
    if (c = f) return 1;
    if (c = f̄) return 0;
    let xᵢ be the top variable of c
    if (c_{xᵢ} = 0) return f_{x̄ᵢ} ↓ c_{x̄ᵢ}
    if (c_{x̄ᵢ} = 0) return f_{xᵢ} ↓ c_{xᵢ}
    return x'ᵢ(f_{xᵢ} ↓ c_{xᵢ}) + x̄'ᵢ(f_{x̄ᵢ} ↓ c_{x̄ᵢ})
}
```

Figure 9.4.   Generalized cofactor.

GC can be used in image computation, as shown in the following theorem, from [CM90].

THEOREM 9.7 Let $F = [f_1, ..., f_n]$ be a vector of Boolean functions, and $R(X)$ a nonempty set. Define GC of f with respect to $R(X)$ as

$$F \downarrow R(X) := [f_1 \downarrow R(X), ..., f_n \downarrow R(X)]$$

Then the image of R under F is equal to the range of the vectorial function $F \downarrow R$, i.e.,

$$Img(R(X), F) = Img(1, F \downarrow R(X)).$$

Now, letting $f_i = x_i$ for $1 \le i \le n$ in the above theorem, we obtain the following corollary:

COROLLARY 9.8 Let X be a set of Boolean variables $\{x_1, ..., x_n\}$, and $R \subseteq \mathbb{B}^n$ a BDD representing a nonempty set. Then R is the range of the vectorial function $[x_1 \downarrow R, ..., x_n \downarrow R]$.

This result is applicable to constraint synthesis in the following sense. Let R be the constraint, and X the set of input variables. Let $X' = [x'_1, ..., x'_n]$ where $x'_i = x_i \downarrow R$. Then the evaluations of X' always satisfy R. Specifically, for any $\alpha \in R$, we have $X'(\alpha) = \alpha$. It follows that X' is a reproductive solution to R. In addition, from the definition of GC, for any input vector α, X' generates a vector in R that is "closest" to α. It is in this sense that GC implements a *least-distance mapping*.

EXAMPLE 9.4 Let us compute the reproductive solutions again for the one-hot constraint

$$x_1 x_2 \bar{x}_3 + x_1 \bar{x}_2 x_3 + \bar{x}_1 x_2 x_3.$$

We solve in the order x_3, x_2 and x_1. Denote the above constraint by c. The following will be used in the computation:

$$c_{x_3} = \bar{x}_1\bar{x}_2,\ c_{\bar{x}_3} = x_1\bar{x}_2 + \bar{x}_1 x_2$$
$$c_{x_3 x_2} = 0,\ c_{x_3\bar{x}_2} = \bar{x}_1$$
$$c_{\bar{x}_3 x_2} = lx_1,\ c_{\bar{x}_3\bar{x}_2} = x_1.$$

Then we have

$$\begin{aligned}
x_3' &= x_3 \downarrow c\\
&= x_3(1 \downarrow c_{x_3}) + \bar{x}_3(0 \downarrow c_{\bar{x}_3})\\
&= x_3
\end{aligned}$$

and

$$\begin{aligned}
x_2' &= x_2 \downarrow c\\
&= x_3(x_2 \downarrow c_{x_3}) + \bar{x}_3(x_2 \downarrow c_{\bar{x}_3})\\
&= x_3(x_2 \downarrow \bar{x}_1\bar{x}_2) + \bar{x}_3(x_2 \downarrow x_1\bar{x}_2 + \bar{x}_1 x_2)\\
&= x_3(0 + 0) + \bar{x}_3(x_2(1 \downarrow \bar{x}_1))\\
&= \bar{x}_3 x_2
\end{aligned}$$

and finally

$$\begin{aligned}
x_1' &= x_1 \downarrow c\\
&= x_3(x_1 \downarrow c_{x_3}) + \bar{x}_3(x_1 \downarrow c_{\bar{x}_3})\\
&= x_3(0 + 0) + \bar{x}_3(x_2(x_1 \downarrow \bar{x}_1) + \bar{x}_2(x_1 \downarrow x_1))\\
&= \bar{x}_3(0 + \bar{x}_2)\\
&= \bar{x}_3\bar{x}_2)
\end{aligned}$$

Comparing this solution to the one given in the example in Section 9.1 (when the state bit is 1), we can see that they are equivalent. To illustrate the least-distance mapping, we add a "distance" column to the table in Figure 9.1, as shown in Figure 9.4. The three legal combinations 001, 010, and 100 are at a distance of 1, 2, and 4, respectively, from the base (000). That mapping of distances 0 to 1, and 5, 6, and 7 to 4 should be clearly least-distance. The combination at distance 3 is equally away from these at 2 and 4. It maps to 2 since the two differ at the variable x_3 that was solved first. This suggests that *Consynth* also gives a least-distance mapping in that earlier example. We will show that indeed the least-distance mapping is also a property of *Consynth*. □

The method based on *Constrain* may produce incorrect results if applied to synthesizing constraints involving state variables. The reason is GC may change the valuation of the state variables in its mapping of a vector of both state and input variables from outside of R to one in R. As a result, what we have generated is an input vector that satisfies R under the *mapped* state but that

distance	p1 p2 p3	x1 x2 x3
0	0 0 0	
1	0 0 1	0 0 1
2	0 1 0	0 1 0
3	0 1 1	
4	1 0 0	1 0 0
5	1 0 1	
6	1 1 0	
7	1 1 1	

may be in conflict with the actual state asserted by the design. A correction to this problem is to solve all the state variables first. However, in general, forcing a variable ordering on a BDD is not feasible.

EXAMPLE 9.5 The constraint $y_1 \to x_1\bar{x}_2 + \bar{x}_1 x_2$, when solved in the order x_1, x_2, and y_1, gives the result $x_1' = x_1$, $x_2' = x_2$ and $y_1' = y_1(x_1\bar{x}_2 + \bar{x}_1 x_2)$. The mapping will reassign the state variable y_1 to maintain the validity of the solution. The vector $y_1 = 1$, $x_1 = 1$, and $x_2 = 1$ is mapped to $y_1' = 0$, $x_1' = 1$, and $x_2' = 1$. In reality, a state assignment is fixed; therefore, we have all three variables assigned to 1, which is incorrect.   □

To conclude the comparison with constraint-based synthesis, we show that our approach achieves the least-distance mapping even when state variables are involved. It also implies that, in the absence of state variables, the two synthesis methods produce identical results.

THEOREM 9.9 The substitution returned by the algorithm in Figure 9.2 maps a vector α to a vector α' in the constraint that has the shortest distance (as given in Formula (9.15)) from α.

PROOF: See Appendix.   □

9.4.1 Randomization

Randomization is strictly controllable in *Simgen* since all solutions are known. However, this is not the case for constraint synthesis. The culprit is the mapping we have just seen. Suppose we try to emulate *Simgen* by assigning biases to the random input variables. If all biases take the value 0.5, *Simgen* would follow a uniform distribution. In *Consynth*, the uniform distribution is only observed at the random input variables. At the output of the mapping, the probability of a vector is the sum of the probabilities of all vectors at the inputs that map to the vector (including the vector itself). In the previous example, the probabilities of 001, 010, and 100 would be 0.25, 0.25, and 0.5, respectively. This distribution is distorted by the mapping and no longer uniform.

However, other forms of randomization are possible. Specifically, the variable-solve-order based randomization (Chapter 3) can be implemented by simply forcing the order on the variable ordering used in constructing the reproductive solution. In addition, weighted distribution (Chapter 3) can be realized by adding new random variables to represent the items in the distribution and give these variables the highest solve order, similar to the procedure in Section 6.5 of Chapter 6.

9.5 Building Circuits from Relations

Kukula and Shiple [KS00] reported a method of building circuits from Boolean relations. The construction takes as input a BDD representation of the relation and generates a gate level netlist. The BDD consists of state and input variables from the design. The netlist is a network of sub-circuits, each of which corresponds to a non-terminal node in the BDD. Topologically, the resulting netlist is isomorphic to the BDD. Inputs to the netlist include the state variables, plus one parametric input for each output of the netlist. The parametric inputs are responsible for generating the randomness at the outputs of the netlist. These outputs are the *reproductive solutions* (Section 9.1) of the input variables in the BDD. They drive the design's inputs.

The sub-circuits are compositions of constructions in three phases. The first phase construction propagates the current state assignment and generates a *weight* signal similar to the *node weight* in *Simgen* (Chapter 6). In the second phase, logics are added to generate an *assignment* signal according to the weights just computed. In the last phase, a *selection* signal is generated that determines whether the assignment signal will be selected to drive an output of the netlist. As can be seen, this construction bears close resemblance to the *Simgen* algorithm. It can be considered an intelligent hardware implementation of the latter. In the sequel, we elaborate on each of these construction phases. We have borrowed terminologies from *Simgen* to facilitate the elaboration.

9.5.1 Computing the Weights

In the first phase, we construct a network of 2-input OR and MUX gates. Each gate corresponds to a BDD node (excluding the ONE and ZERO nodes). An OR gate corresponds to an input node (the node of an input signal) and a MUX gate corresponds to a state node. The select line of a MUX gate connects to the corresponding state signal. The inputs of an OR or MUX gate, corresponding to a node n, connect to the outputs of the two gates corresponding to the two children of n. The ONE and ZERO nodes become the constants POWER and GROUND, respectively.

After constants are propagated, the output of these OR gates represents the weight of their respective gates. Unlike *Simgen*, where weights are real num-

bers, the weight here is either 0 or 1. The output of the gate corresponding
to the root of the BDD indicates the satisfiability of the following relation: 1
means satisfied and 0 means a conflict. When it is not a conflict, the weights
are used to generate the input assignment in the second phase.

EXAMPLE 9.6 Figure 9.5 illustrates weight computation for constraint $y_1 x_1 x_2 +$
$y_1 \overline{x}_1 y_2 + \overline{y}_1 y_2$, where x_1 and x_2 are the input variables, and y_1 and y_2 the state
variables. The left-hand side of the figure is the BDD of the constraint (the
solid and dotted arrows represent the *then* and *else* branches, respectively). The
right-hand side of the figure shows the weight computing circuit consisting of
MUXes for state variables and OR gates for the inputs.   □

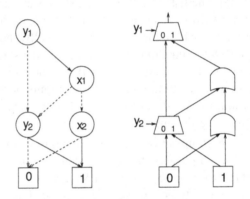

Figure 9.5.   Example: weight computation circuit.

9.5.2   Computing the Assignments

In the second phase, we combine the weights and parameter inputs to produce
an assignment signal for each node. Essentially, we add a MUX gate for each
input node in the BDD.

Recall that to satisfy the constraint (relation), a node branches to another node
only if the weight of the second node is positive. Consider a non-terminal BDD
node n. The weights of its *then* and *else* children have four combinations: $\{0,0\}$,
$\{0,1\}$, $\{1,0\}$, $\{1,1\}$. The first combination occurs only when the constraint is
not satisfiable or the to-be-computed assignment is not going to be selected in
phase three (see Theorem 6.8). In either case, the combination is a don't care
situation. In the second and third combinations, we are forced to assign 0 (resp.
1) since the only branch with a positive weight is the *else* (resp. *then*) branch.
In the last combination, both 0 and 1 are possible. This is the time to use a
random parametric variable as the assignment. It should be noted that the above

analysis, based on the weight computation concept from *Simgen*, is also key to the *Consyth* algorithm discussed in Section 9.2.1.

The MUX gate that generates the assignment for an input node is derived as follows. Let p be the parametric variable and x the input variable to be assigned. Let t and e be the weight of the *then* and *else* branches, respectively. Let p be assigned to 0 in the don't care combination 0,0 of the weights. Then the assignment to x can be written as (? is the ITE operator):

$$x = \bar{t} \,?\, 0 : (\bar{e} \,?\, 1 : p) = e \,?\, p : t$$

which is a MUX with e as the select signal, and p, t as the 1 and 0 inputs, respectively.

9.5.3   Selecting the Outputs

So far, the sub-circuit consists of the weight logic for all nodes and the assignment logic for all input nodes. We now need to find an assignment for each input variable in the constraint. If the weight of the root node is 0, then the relation is not satisfied and the assignments are don't cares. Otherwise, out of the assignments made from all the input nodes, exactly one is selected for each input variable as the final assignment to the corresponding output of the constructed netlist.

To do this, we need to find a path from the root to the terminal node ONE, as in the *Walk* procedure (Section 6.2.2), so that only the nodes on the path are selected for the final assignments. The selection is defined recursively: a node is selected iff one of its parents is selected and then branches to that node.

Therefore, a node is responsible for generating the selection signals to be used by its two children. If the node itself is not selected, then both selections are 0. Otherwise, the *then* selection equals the assignment made at the node, and the *else* selection equals the negation of that assignment. In addition, each node also generates a selection signal for itself that is the OR of selection signals from all of its parents. Note that at most one of these selection signals can be 1. A node is selected iff its own selection signal is 1. A special case is the root node, which has one incoming selection signal that is tied to constant 1.

Finally, the assignment to an input variable is a bus driven by all the nodes of the variable. Each driver is a tri-state buffer whose input and enable are the assignment and selection signal of the corresponding input node, respectively. Figure 9.6 illustrates the complete logic of sub-circuits for the input and state nodes. Figure 9.6 gives the netlist for the constraint example in Figure 9.5.

9.6   Synthesis using SAT solving

The constraint synthesis algorithm given in Figure 9.3 has in mind a BDD implementation. Although a compact representation of Boolean functions, BDD does face the problem of a size blowup when the functions are too compli-

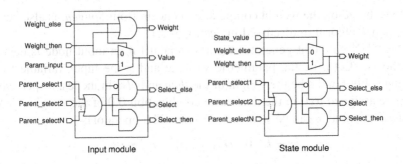

Figure 9.6.   Input and state modules.

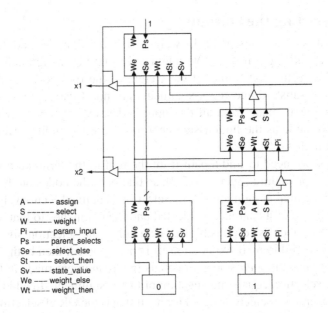

Figure 9.7.   Netlist representing a constraint.

cated. In addition, building the conjunction of BDDs, required when multiple constraints must be solved together, aggravates the problem.

Alternatively, *Consynth* can be implemented with techniques used in SAT-based image computation [McM02, GGA04]. Key to the synthesis algorithm is the computation of the projections (Section 9.2.1), which essentially is a sequence of existential quantifications. We give a SAT-based constraint synthesis algorithm by focusing on this computation. We omit the optimizations from variable removal and *BDD_Restrict*.

The main idea in SAT-based image computation is quantification through enumeration. Let $f(X, Y)$ be a function over two sets of variables X and Y, where Y are to be existentially quantified out. Let $g(X)$ be the quantification result. In [McM02], $g(X)$ is obtained by repeatedly solving $f \wedge \bar{g}$ until it is unsatisfiable. From every solution obtained, the Y assignment is ignored, and the X assignment is reduced as much as possible, resulting in a maximum *blocking* cube. This cube is added (disjunctively) to g before the next SAT solving. This idea is improved in [GGA04], where the blocking cubes are replaced by cofactors of f: after solving $f \wedge \bar{g}$, the cofactor of f with respect to the Y assignment of the solution is added to g. The intuition is that each Y assignment may correspond to multiple X cubes, including the blocking cube. Therefore, fewer enumerations are needed. In addition, the functions f and g are represented as OR-INVERTER circuits with a multi-level hashing, which are suitable for efficient cofactoring and size reduction. The basic quantification algorithm is shown in Figure 9.8.

```
// g is initially empty.

sat_ex_quant(f, X, Y, g) {
    if SAT(f ∧ ḡ) fails return g;
    let s be the solution over Y;
    let fs be the cofactor;
    return sat_ex_quant(f, X, Y, g ∨ fs);
}
```

Figure 9.8. SAT-based quantification.

It would be problematic to simply quantify out one variable in each recursive call by replacing $\exists_x f$ in Figure 9.3 with the above SAT quantification, and pass on the result as the constraint f to be used in the next recursive call. This would run the risk of doubling the size of f each time, leading to a simple-minded exponential expansion of f. Instead, we accumulate the quantification variables, all of which are quantified out in each recursive call from the original f. This SAT-based constraint synthesis algorithm is given in Figure 9.9.

9.7 Experimental Results

We implemented the proposed algorithm, as well as two related methods, for evaluation. The examples with constraints are from actual designs. Each example has up to a few hundred constraints that involve anywhere from hundreds to over a thousand input and state variables. The actual numbers of variables in the examples range from a few to over ten thousand.

$Sat\_Consynth(f, X, Y)$ {
   if (X contains no input variables) **return**;
   $g = sat\_ex\_quant(f, X, Y, \emptyset)$;
   let x be an input variable in X
   $\sigma_x = x\, g_x + \overline{g_{\overline{x}}}$;
   $X = X \setminus \{x\}$;
   $Y = Y \cup \{x\}$;
   $Sat\_Consynth(f, X, Y)$;
}

Figure 9.9.   SAT-based Constrain synthesis.

In Table 9.1, we compare the construction time (in seconds) and the size of the BDDs (in the number of BDD nodes) representing solutions from three methods: *Simgen* – the vector generation tool from [YSP$^+$99], *BU* – the Boolean Unification method from [MN89], and *Consynth*, the method presented in this chapter. All three share the same flow, including design compilation and constraint partitioning, up to the construction of BDDs representing the constraints. *Simgen* ends after this stage, while *BU* and *Consynth* starts their respective constraint solving processes. As can be seen, in *Consynth*, the extra time spent in constraint solving is modest compared to the time for building the constraint BDDs, indicated by the time used by Simgen. Also, the final BDD size is reduced in all but one case. This should also apply to the comparison of our method versus the method in [KS00], which would produce BDDs of the same size as those produced by *Simgen*. The classic *BU* approach, with recursive substitution and lacking the optimizations proposed, tends to generate larger BDDs, and actually in half of the cases timed out (t/o) after thirty minutes.

The last column in the table reports the result of variable removal, where an entry m/n means m out of n input variables are removed from the final solution.

In Table 9.2, we report the results of using our method in vector generation by comparison with *Simgen*. The first two columns give the time (in seconds) spent by each method on vector generation in simulation. Each simulation runs for 10,000 clock cycles. The third column reports the speedup using *Consynth*. We have already seen the reduction of BDD size in *Consynth*. However, this is not the main reason for the order of magnitude improvement of the vector generation speed. When used as an explicit vector generator, *Consynth* uses time linear in the *number of variables* in the solution, whereas *Simgen* uses

Table 9.1. Complexity of Solutions.

circuit	Simgen		BU		Consynth		
	time	size	time	size	time	size	removed
des1	19.0	2105	t/o	–	29.0	3814	21/115
des2	27.0	2345	34.0	1599	30.0	822	238/423
des3	12.0	2437	14.0	1347	14.0	1217	6/174
des4	63.0	2847	88.0	15675	69.0	1521	11/134
des5	15.0	4679	92.0	16056	40.0	1641	15/63
des6	73.0	21822	t/o	–	81.0	8108	132/689
des7	237.0	28430	t/o	–	279.0	18480	23/207
des8	141.0	53788	t/o	–	169.0	9968	45/283

time linear in the *number of BDD nodes*.[1] This contrast is especially noticeable when the BDD size gets large in both methods.

Table 9.2. Speedup of Vector Generation.

circuit	Simgen	Consynth	speedup
des1	71.06	7.30	9.73
des2	34.66	11.68	2.97
des3	22.35	10.98	2.03
des4	14.70	2.18	6.74
des5	10.96	5.56	1.97
des6	181.72	6.74	26.96
des7	124.61	14.57	8.55
des8	219.4	9.53	23.02

9.8 Summary and Discussion

We have described a method of synthesizing Boolean constraints. Using constraints is an effective alternative to writing complicated environment models in functional verification. Constraint synthesis facilitates the application of

---

[1] Since state variables are assigned by the design, one cannot walk randomly down the BDD to generate vectors. Instead, a bottom-up *weight* computation (in time linear to the BDD size) has to be performed to guide the walk [YSP+99].

this methodology to areas where no proprietary constraint solving capability is available, e.g., model checking and emulation.

We have also provided a survey of constraint synthesis methods that are known to us. Although Boolean Unification is shown to be *unitary* [W. 87], that is, if there exists a solution, then there is a unique *mgu* from which all solutions can be derived as its instances, the uniqueness is nevertheless only up to an equivalence relation [Sie84]. For example, a BU problem can have two *equivalent mgus* which can be derived from each other. The fact that there actually can be multiple, functionally distinct *mgus* gives rise to the need for optimization. Our application of don't cares in cascaded synthesis is such an attempt. It is also obvious that synthesis results vary with variable ordering in both the BU based on the *constrain* operation and the one built upon Boole's method. In the latter case, Büttner [W. 88] has shown how to decide the variable ordering in order to obtain an *mgu* with the minimal number of variables.

An approach that allows maximal optimization should be one in which both the don't cares and variable ordering are exploited. In general, we do not have to follow any particular order in the synthesis, e.g., we can solve variable x before variable y on one path of the decision making, and in the reverse order on another. This would lead us to depart from the class of least-distance mappings, which is a sufficient but not necessary condition for constrained vector generation. All we require here is a general solution that encompasses all of the ground solutions.

Chapter 10

CONSTRAINT DIAGNOSIS

The diagnosis of constraints is a practical concern in many fields, e.g., in artificial intelligence [Dav84] and constraint programming [ABC+97]. A constraint conflict stalls the progress in simulation and leads to vacuous proofs in formal verification. Although manifested as a constraint solving problem, constraint conflicts may very well originate in abnormalities in the design or in the interaction between the design and the constraints. In fact, many constraint conflicts take place when the design enters some unexpected state. For example, the constraints $a \rightarrow c$ and $b \rightarrow \bar{c}$ are in conflict if and when the state $a \wedge b$ is reached. A first possibility is that this is a genuine *over-constraint* case: the constraint developer should have allowed some input assignment for this state. On the other hand, she might have assumed that this state is never reachable — a scenario common to cooperative state machines that are supposed to have mutually exclusive states. In this case, the constraint conflict indicates the violation of a valid assumption, thus revealing a design error. However, to complicate matters, the error itself may be attributed to a missed constraint or constraints that are too loose to prevent the design from entering the erroneous state. This latter situation characterizes the *false-negative* results or false alarms due to *under-constraints* in verification. Whether or not a design error is caused by an under-constraint must be determined in the normal design debugging process. Actually, incrementally adding in missing constraints is a common practice in constraint-based verification.

In this chapter, we will focus on the diagnosis of over-constraints through static analysis. We will deal with propositional constraints only, since most sequential constraints in practice are equivalent to state machines with some propositional constraints asserting on their states, as we have seen in Chapter 4.

10.1 The Illegal States

Recall in the test generation algorithm discussed in Chapter 6 that if the root of the constraint BDD has a weight of zero, then no satisfying input assignment can be made, or equivalently, the design has entered an illegal state. This gives the definition of illegal states: they are the states under which the constraints are not satisfiable. Therefore, over-constraint is possible only if there are illegal states.

Let $f(X, Y)$ be a constraint over the set of input variables X and state variables Y. The set of illegal states with respect to f, denoted by $\psi(f)$, is computed as:

$$\psi(f) = \forall_X(\overline{f}). \tag{10.1}$$

Note that, for a constraint defined over only input variables, the set of illegal states is either empty (if the constraint is unsatisfiable) or a tautology (if the constraint is satisfiable).

The following two results concern the determination of the illegal states of a conjunction of constraints. First, an illegal state of a constraint f_i is also an illegal state of a conjunction of constraints that include f_i.

LEMMA 10.1 Let $g = f_1 \wedge f_2 \wedge \cdots \wedge f_n$, then for all $1 \le i \le n$

$$\psi(f_i) \le psi(g).$$

PROOF: $psi(\bigwedge_{j=1}^{n} f_j) = \forall_X(\overline{f_1} + \ldots + \overline{f_n}) \ge \forall_X(\overline{f_i}), 1 \le i \le n.$ □

Second, given the disjoint-input-support partition of a set of constraints, the illegal states of the conjunction of these constraints is the union of the illegal states of the components in the partition.

LEMMA 10.2 Let $g = f_1 \wedge f_2 \wedge \cdots \wedge f_n$ where $f_1, \ldots, f_n$ have disjoint-input-supports, then

$$\psi(g) = \bigvee_{i=1}^{n} \psi(f_i).$$

PROOF: Let $X_1, \ldots, X_n$ be the partition of X corresponding to $f_1, \ldots, f_n$, then

$$
\begin{aligned}
\psi(\bigwedge_{j=1}^{n} f_j) &= \forall_X(\overline{f_1} + \ldots + \overline{f_n}) \\
&= \forall_{X_1}(\overline{f_1}) + \ldots + \forall_{X_n}(\overline{f_n}) \\
&= \forall_X(\overline{f_1}) + \ldots + \forall_X(\overline{f_n}) \\
&= \psi(f_1) + \ldots + \psi(f_n)
\end{aligned}
$$

☐

The above illegal states computation can be done with the BDD quantification operations, or with SAT, using the enumerative method discussed in Section 9.6.

If there are no illegal states, then there are no over-constraints, and the constraint analysis is completed. If there are illegal states, however, we need to determine if they are reachable in the constrained design. An illegal state can be unreachable already in the unconstrained design, or become unreachable because of the constraints. In fact, the second scenario serves the main purpose of having constraints: to make some "bad" states unreachable.

10.2 Reachability Analysis

Given a set of illegal states, p, whether or not an over-constraint will ever happen can be determined by a slightly modified BDD-based reachability analysis (Chapter 5, Figure 5.7). The modification adds an intersection check between p and the image. If they intersect, then some illegal state is reachable, and a trace is generated. No illegal states are reachable if the reachability analysis finishes and no intersection is detected.

This process is very close to the reachability analysis of a safety property, with one subtle distinction: in the latter, constraints are used to reduce the image before the intersection check to avoid false negativity. This is not necessary in the former.

With this distinction in mind, we can apply other static verification methods to the our problem. The most scalable is the SAT-based bounded model checking. Bounded model checking can discover property violations, that is, in our case, the confirmed reachability of illegal states. In addition, bounded model checking can be augmented with techniques, such as induction, to conclude the absence of property violations.

10.3 Locating the Conflict Source

Once a reachable illegal state is found, we further diagnose the conflict by finding a small set of constraints that are actually responsible for the conflict.

Because the illegal states were computed by summing up the illegal states from each group in the constraint partition (Lemma 10.2), we first localize the search to the groups that produced the reached illegal state. Next, we simplify the constraints in the selected groups with the reached illegal state. If a constraint is reduced to the constant zero, then we do not need to look further. We return all such constraints, each as a sufficient reason for the illegal state (thus a conflict). Otherwise, we need to find a minimal conflicting set of constraints for each selected group.

In BDDs, a minimal conflicting set can be obtained by forming the conjunction of subsets of the constraints in increasing cardinality, until an empty conjunction is found. This can be expensive if the number of constraints is large. Alternatively, a minimal conflicting set can be approximated by computing the unsatisfiable core (Chapter 5) using SAT. We form the SAT problem by conjoining the simplified constraints. The problem must be unsatisfiable and therefore, a core can be constructed. The constraints used in the core are returned as the subset responsible for the illegal state.

10.4   Fixing Over-constraints via Illegal States Removal

As we have seen in the introduction, reaching an illegal state can be attributed to many reasons: a genuine over-constraint, a design error, or an under-constraint. It is up to the designer or the constraint developer to determine which reason it is. If it is an over-constraint, then a subset of conflicting constraints can be computed, as in the previous section. Furthermore, the over-constraint can be fixed by methodologically "eliminating" the illegal states from the constraints.

From Lemmas 10.1 and 10.2, if we eliminate the illegal states from each constraint, and from each component of the partition, then the resulting constraints are free of illegal states. The elimination is carried out as a constraint relaxation, as stated in the following lemma.

LEMMA 10.3 Let f be a constraint, d a subset of the set of illegal states, $\psi(f)$, and X the set of input variables. Then

1. $\psi(f + c) \leq \psi(f)$, and

2. $\psi(f) - \psi(f + c) = d$ for all c such that $\exists_X(c) = d$.

PROOF: Since

$$\begin{aligned}
\psi(f + c) &= \forall_X(\overline{f + c}) \\
&= \forall_X(\overline{f}) \cdot \forall_X(\overline{c}) \\
&= \psi(f) \cdot \forall_X(\overline{c})
\end{aligned}$$

therefore, $\psi(f + c) \leq \psi(f)$, and

$$\begin{aligned}
\psi(f) - \psi(f + c) &= \psi(f) \cdot \overline{\psi(f) \cdot \forall_X(\overline{c})} \\
&= \psi(f) \cdot \left(\overline{\psi(f)} + \exists_X(c) \right) \\
&= \psi(f) \cdot d \\
&= d
\end{aligned}$$

$\square$

The above lemma shows that a subset d of the illegal states of f can be removed by relaxing f to $f + c$, where c allows some input assignments for every state in d. Furthermore, the relaxation does not introduce new illegal states.

10.5   Summary

We have described a constraint diagnosis method based on the analysis of illegal states. As we have previously stated, one advantage of constraints over traditional testbenches is that the former can be methodologically analyzed and debugged. This is evidenced in the formulation of the constraint diagnosis problem and in the application of formal verification techniques.

Chapter 11

WORD-LEVEL CONSTRAINT SOLVING

In this chapter, we provide a survey of recent developments in word-level constraint solving. Previously, we have discussed constraint solving problems in terms of their Boolean formulation, mainly using BDDs. We also discussed converting an RTL, or a word-level description to a bit-level netlist which is amenable to Boolean analysis. However, the bit-level representation, when compared to its word-level origin, may have two disadvantages: one, it can be much larger than the original word-level representation, and two, constraint solving itself may be less efficient. Word-level models deal with data types that are generally representable with integers of finite domains. Although it is still an open question as to when bit-level modeling is more appropriate for constraint solving than word-level modeling or vice versa, there exist empirical data indicating that modeling integers at word-level can be advantageous.

The overhead of coercing word-level constraint solving to Boolean satisfiability is quite noticeable, especially in the case of BDD-based satisfiability. BDDs are desirable representations: all solutions are efficiently accessible and randomization strictly controllable. However, a constant effort must be made to keep BDDs small enough to realize these advantages. We have seen that techniques such as partitioning, extraction, and decomposition have worked well enough to enable a BDD-based constraint solver to be of practical use in industrial settings. However, for data intensive designs such as data-paths and DSP, BDD size blowup is almost unavoidable. In particular, BDDs representing the high order output bits of an multiplier have an exponential size regardless of their variable orderings [LS81, Bry91b]. To address this deficiency, other types of decision diagrams have been developed. For example, Multi-Terminal Decision Diagrams (MTDDs) [CMZ+93] and Algebraic Decision Diagrams (ADDs) [BFG+93b] are variations of BDDs that map Booleans to integers. Binary Moment Diagrams (BMDs) [BC95] use a moment-based decomposition function that map Boolean vectors to integers. BMDs, being more compact in representing arithmetic logics, enabled the first symbolic verification of multipliers with a word-width of up to 256 bits. Later, Hybrid Decision Diagrams (HDDs) [CFZ95] combined BMDs and MTDDs so that control and arithmetic logics were represented in their respective Boolean and integer domains. Multiple-valued extensions to BDDs have also been studied, for example, in [SKMB90, LL92, Mil93, DBR96]. Intrinsically, however, the succinctness of any decision diagram still draws heavily upon the decomposition type and variable ordering. Furthermore, special care is required when both bit- and word-level logic are present.

SAT and ATPG are constraint solving approaches complementary to BDDs. They are not symbolic data structures that represent all solutions. Instead, they solve the problem by trying to derive a solution one at a time. This categorizes SAT and ATPG, when applied to test vector generation, as on-line, as opposed to the BDDs, which are off-line, because the real solving is done prior to simulation. ATPG algorithms work directly off a bit-level netlist, whereas most SAT solvers take as input a CNF formula that is linear in the size of a bit-level netlist. However, the netlist itself may be much larger than the original representation of the problem. For example, the number of two-input gates taken to implement a combinational multiplier is quadratic in the width of the operants. One way to solve this problem is to reduce the signal width, as was done by Johannsen and Drechsler [JD01]. Other attempts at word-level constraint solving include extensions of logic-based methods such as SAT, the classical linear and general constraint and logic programming solutions, or a combination of both.

Linear Programming (LP) originated in the Operation Research field for solving optimization problems over reals. LP algorithms, such as the Simplex and interior-point methods, optimize a cost function for a given system of linear constraints over reals. Integer Linear Programming (ILP) is the counterpart of LP, which finds solutions in integers. Mixed real and integer linear programming solves problems in both the real and integer domains. A special kind of ILP, called 01-Integer-Linear-Programming (01-ILP), finds solutions in Boolean space. Traditionally, ILP problems are solved in a general branch-and-bound framework where LP-relaxation, i.e., finding a solution by relaxing the integral constraint on the variables, plays a central role in providing bounds on solutions. The branch-and-bound framework later evolved into branch-and-cut and other variations which essentially prune the choices of branching. For example, the earliest branch-and-cut scheme adds the Gomory-Chvátal *cutting-plane* (CP) inequalities as additional constraints to restrict branching.

Constraint Logic Programming (CLP) combines constraint solving and logic programming. CLP is a methodology rather than a technology that can be clearly defined. A CLP tool may solve problems ranging from linear and non-linear constraints, to propositional logics, and even temporal logic. The underlying constraint solvers can be logic-based (e.g., SAT), mixed LP and ILP, and interval arithmetic, over the real, finite, and Boolean domains.

There is extensive literature on the general constraint solving paradigm briefly mentioned above. For example, Jaffar and Mahr [JM94] give an excellent survey on CLP. In this chapter, we will focus on the recent developments of the application and the extension of general constraint solving techniques in the EDA field, particularly constraint solving at the word-level and mixed bit-and-word-level.

One of the early developments is the adoption of DPLL-based SAT algorithms in 01-ILP. Hooker [Hoo94] relates cutting-planes in 01-ILP to logical inferences and points out that LP-relaxation can be replaced by discrete relaxation. He also shows an extension of resolution to *multivalent clauses*, i.e., clauses with multi-valued variables. Barth [Bar95a] demonstrates a logic-based 01-ILP optimization method using a modified Davis-Putnam algorithm. This method compares well with traditional ILP approaches. Aloul *et al.* [ARMS02], Chai and Kuehlmann [CK03], and Sheini and Sakallah [SS05] extend the idea by applying the latest SAT techniques, including decision heuristic, clause learning, and efficient BCP, showing their advantages in problem formulation and performance over CNF-based SAT in several EDA optimization applications. Liu and Kuehlmann [LK03] extend the same SAT techniques to multivalent clauses and show improved performance over CNF-based SAT in logic synthesis problems.

There are also pure integer and logic programming based approaches. For instance, Zeng *et al.* in [ZKC01] and [ZCR01] model both the control and arithmetic logics as ILPs and CLPs, respectively, in functional test generation. Brinkmann and Drechsler [BD02] convert RTL data-path verification problems to ILPs. More recently, hybrid approaches have been advocated because of the realization that both Boolean SAT/ATPG and constraint programming have their own domain-specific strengths. In particular, it is highly desirable to leverage the greatly-enhanced SAT solving techniques developed in recent years in a mixed bit- and word-level constraint solving environment. Fallah *et al.* [FDK98] were among the first to apply SAT and ILP to functional vector generation. Huang and Cheng [HC00] and Iyer [Iye03] address constraint solving in assertion checking with ATPG and ILP in the tool RACE. Parthasarathy *et al.* in [PICW04] and Iyer *et al.* in [IPC05] address a hybrid SAT and ILP approach to constraint solving for RTL circuits in the tool HDPLL. They apply such SAT techniques as conflict analysis, lazy evaluation, and propagation to mixed bit- and word-level constraints. Parthasarathy *et al.* in [PICB05] extend their earlier work by employing an ATPG-based decision heuristic. Audemard *et al.* in [ABC$^+$02] and Bozzano *et al.* in [BBC$^+$05] present a tool called MATHSAT that solves Boolean, real, and integer constraints by combining SAT and various

LP and ILP solvers. The constraints involving real numbers are used in, for example, the bounded model checking of timed systems [ACKS02].

We should also mention the tools UCLID [SLB03], ICS [FORS01, dMRS02], and CVC Lite [BB04]. Each of these tools takes as input a quantifier-free first-order logic formula and decide its validity or satisfiability. Special predicates, such as ones on arithmetic and memory operations, can be handled with interpreted and uninterpreted functions and constant symbols. UCLID converts the input formula to a propositional formula through several steps of translation and optimization. The result is then sent to a BDD or SAT solver for validity checks. ICS and CVC Lite instead decide the propositional portion of the input formula using SAT, and the rest with decision procedures (and variants) due to Shostak [Sho84] and Nelson and Oppen [NO79]. In CVC Lite, SAT and the other decision procedure are executed in the same loop, and thus decisions in the former are immediately validated by the latter. This is referred to as the "eager" approach to combining the two decision procedures (MATHSAT and HDPLL take the same approach). ICS, on the other hand, takes the "lazy" approach: the SAT problem is solved before the other decision procedure is invoked.

In the remainder of this chapter, we present the main ideas from these approaches for using SAT, ATPG, ILP, and their combinations in a mixed bit- and word-level constraint solving for RTL circuits. We will discuss SAT-based approaches to 01-ILP, multi-valued logic, and word-level constraint solving. We will also address the modeling of RTL circuits and constraints as integer linear constraints.

11.1   DPLL-based 01-ILP

As reviewed in Chapter 5, the main features of modern DPLL-based SAT solvers are conflict analysis that drives clause learning and backtracking, efficient implication and constraint propagation, and decision heuristics influenced by conflict analysis. A high-level DPLL schema is given in Figure 11.1. Conflict analysis performs a sequence of resolutions in the implication graph and finds a set of variable assignments responsible for the conflict. Therefore, developing a DPLL-based 01-ILP solver amounts to defining resolution, implication and constraint propagation in 01-ILP. There are other SAT features that we omit in this discussion. They are important, but irrelevant to the extension to 01-ILP. For example, we omitted discussion of random restart, which is independent of problem representation, and clause minimization, which corresponds to such constraint preprocessing techniques as the ones used in [Pug92]. The discussion in this section is mainly based on the results from [Hoo94, Bar95a, ARMS02, CK03].

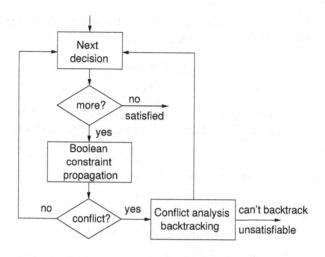

Figure 11.1. A DPLL-based SAT solver.

11.1.1 Linear Pseudo Boolean Constraints

01-ILP deals with Linear Pseudo Boolean (LPB) constraints. An LPB is an inequality of the form:

$$\sum_{i=1}^{n} a_i x_i \geq b, \quad a_i, b \in \mathbb{Z}, \text{and } x_i \in \{0, 1\}. \tag{11.1}$$

All non-tautological LPB constraints can be converted through arithmetic equivalence translation and coefficient reduction [CJP83] to a normal form, where the integer coefficients and the left-hand-side (LHS) of the inequality satisfy

$$b \geq a_n \geq \ldots \geq a_1 \geq 1, \text{and } x_i \neq x_j \text{for } i \neq j. \tag{11.2}$$

Normalization can be divided into two phases. In the first phase, product terms are merged and coefficients made positive by utilizing the equality $x_i = 1 - \overline{x_i}$. For example,

$$
\begin{array}{rcl}
8x_1 + 8\overline{x}_1 - 3\overline{x}_2 - \overline{f}x_3 - 2x_3 & \geq & b \\
\equiv \quad 8 - 3(1 - x_2) - \overline{x}_3 - 2(1 - \overline{x}_3) & \geq & b \\
\equiv \quad 3x_2 + \overline{x}_3 & \geq & b - 3
\end{array}
$$

It is clear that the above constraint is not a tautology iff $b - 3 \geq 1$. All tautological constraints are ignored. The second phase applies coefficient reduction to ensure that no coefficient is greater than the new right-hand-side (RHS) constant, b'. This is done by simply replacing all coefficients greater than b'

with b'. Note that this transformation preserves the satisfiability of the original constraint. For example, if $b' = 2$, then

$$3x_2 + \overline{x}_3 \geq 2$$
$$\equiv \quad 2x_2 + \overline{x}_3 \geq 2$$

A bit-level netlist corresponds to a set of normalized LPB constraints, which are direct translations of CNF clauses. For example, the three clauses for an AND gate $a = b \ \& \ c$:

$$(\overline{a} + b), \ (\overline{a} + c), \ (a + \overline{b} + \overline{c})$$

translate to three LPB constraints:

$$\overline{a} + b \geq 1, \ \overline{a} + c \geq 1, \ a + \overline{b} + \overline{c} \geq 1.$$

However, a bit-level netlist may not be the best source to derive LPB constraints from. In fact, a main motivation for the LPB solver developed in [ARMS02] is to avoid the large bit-level representation of certain EDA optimization problems, and derive LPB constraints directly from the problems. For instance, a straightforward bit-level netlist of a counting constraint which chooses k out of n objects would yield $\binom{n}{k+1}$ $k + 1$-literal clauses [ARMS02]. Although an n-bit adder followed by a comparison with k gives a linear implementation of this constraint, it is still less compact than the constraint $x_1 + \ldots + x_n \leq k$, which normalizes to $\overline{x}_1 + \ldots + \overline{x}_n \geq n - k$. Many other optimization problems, including Max-SAT, Max-Ones, and minimal cover, have similar direct representation in LPB constraints.

11.1.2   Cutting-planes in ILP

Resolution is a complete decision procedure for propositional logic in a clausal form (a propositional formula corresponds to a combinational circuit, which easily transforms to a set of clauses). The *unit resolution* method used in the Davis-Putnam algorithm repeatedly selects a variable to resolve and eventually reaches an empty clause or a satisfying assignment. Parallel in LP, a complete decision procedure, the *Fourier-Motzkin Elimination* (FME) [Wil76], also removes variables by applying the "non-negative linear combination" rule of inequalities:

$$\frac{\lambda_1(a_1x_1 + \ldots + a_nx_n \geq b)}{\lambda_2(c_1x_1 + \ldots + c_nx_n \geq d)} \frac{}{(\lambda_1a_1 + \lambda_2c_1)x_1 + \ldots + (\lambda_1a_n + \lambda_2c_n)x_n \geq \lambda_1b + \lambda_2d}, \ \lambda_1, \lambda_2 \geq 0. \tag{11.3}$$

Note the following correspondences between clausal resolution and FME. An "empty clause" in FME can be any unsatisfiable inequalities such as $-1 \geq$

0. Similar to clausal resolution, FME only applies to a variable x_i and two inequalities in which the coefficients of x_i, say a_i and c_i, have opposite signs, so that $\lambda_1 a_i + \lambda_2 c_i = 0$ and $\lambda_1, \lambda_2 \geq 0$ have a solution. For any variable x_i whose coefficients are always positive or always negative, that is, x_i is one-side unbounded, the inequalities that contain x_i can be dropped since they can always be satisfied by assigning a large enough value to x_i. The one-side unbounded variables correspond to the pure literals (Section 5.6.2, Chapter 5) in clausal resolution.

However, FME does not directly extend to ILP as a decision procedure. FME *weakens* the information contained in the original inequalities [Wil76], sometimes in a drastic way. For example, resolving on x_1 and x_2 on the four clauses, $(x_1 + x_2)$, $(x_1 + \overline{x}_2)$, $(\overline{x}_1 + x_2)$, and $(\overline{x}_1 + \overline{x}_2)$, gives an empty clause. However, if we eliminate x_1 and x_2 from the corresponding LPB constraints

$$
\begin{array}{lrcl}
(1) & x_1 + x_2 & \geq & 1 \\
(2) & x_1 + \overline{x}_2 & \geq & 1 \\
(3) & \overline{x}_1 + x_2 & \geq & 1 \\
(4) & \overline{x}_1 + \overline{x}_2 & \geq & 1
\end{array}
$$

by a) adding (1) and (3) which gives $2x_2 \geq 1$ (using $\overline{x}_1 = 1 - x_1$), b) adding (2) and (4) which gives $-2x_2 \geq -1$ (note that the other two additions result in tautologies), and c) adding the results of a) and b), we would get the tautology $0 \geq 0$. This obviously contradicts the fact that the original clauses and the LPB constraints are conflicting. However, FME does give the correct answer in the sense that the LPB constraints, if treated as LP inequalities, are satisfiable. In fact, the solution would be $x_1 = x_2 = 0.5$.[1]

Therefore, the root cause is that the implicit integral constraints in ILP are not realizable in FME. This can be fixed by *rounding*, a key idea in CP. CP consists of two rules, the non-negative linear combination rule given in (11.3), and the rounding rules:

$$
\frac{a_1 x_1 + \ldots a_n x_n \geq b}{\frac{a_1}{\lambda} x_1 + \ldots + \frac{a_n}{\lambda} x_n \geq \lceil \frac{b}{\lambda} \rceil}, \quad \frac{a_i}{\lambda} \in \mathbb{Z} \tag{11.4}
$$

and

$$
\frac{a_1 x_1 + \ldots a_n x_n \geq b}{\lceil \frac{a_1}{\lambda} \rceil x_1 + \ldots + \lceil \frac{a_n}{\lambda} \rceil x_n \geq \lceil \frac{b}{\lambda} \rceil}, \quad x_i \geq 0 \tag{11.5}
$$

As before, λ is a non-negative real. Note that in the first rounding rule, rounding happens only on the RHS of the constraint, since a_i must be evenly divided by λ. In the second rounding rule, rounding happens on both the LHS and RHS,

---

[1] To use FME as a solution method, one eliminates all but one variable, then propagates the solution from the one-variable inequality to inequalities with two variables and so on, to obtain the solution to all the variables.

provided that the variables $x_1, \ldots, x_n$ are non-negative. Therefore, this rule applies to special ILP problems such as 01-ILP.

Schrijver [Sch86] has shown that the CP rules form a complete proof system for ILP. It was proven by Cook *et al.* [CT87] that CP polynomially simulates and can be exponentially stronger than resolution if the clauses are converted to ILP inequalities. For instance, the unsatisfiability proof of the *pigeon-hole* problem (fit n pigeons into $n-1$ holes) requires n^3 CP steps but exponentially many resolution steps [CT87].

Let us return to the previous example. A conflict can be derived from the two intermediate results $2x_2 \geq 1$ and $-2x_2 \geq -1$ as follows:

$$
\begin{array}{rcll}
2x_2 & \geq & 1 & \\
x_2 & \geq & 1 & \text{(divide by 2 and round up)} \\
2x_2 & \geq & 2 & \text{(multiply by 2)} \\
-2x_2 & \geq & -1 & \text{(add)} \\
\hline
0 & \geq & 1 &
\end{array}
$$

11.1.3   A DPLL-based 01-ILP Algorithm

ILP inequalities are more expressive than clauses, and because of this the CP rules need to be more flexible. One of the consequences of this flexibility is that multiple variables can be eliminated in one step. All of these contribute to the succinctness of CP proofs. However, the same flexibility also makes it difficult to find the proofs. For example, it is not clear how to choose the weights when applying the non-negative linear combination rule. Also, the search space in which CP tries to find implied variable dependencies relevant to the proof grows rapidly with the size of the variables' domains. Therefore, CP-based approaches are often only applied to 01-ILP problems [BB95]. In this subsection, we examine a 01-ILP approach, mainly based on the one by Chai and Kuehlmann [CK03], that adopts techniques from modern DPLL-based SAT solvers. The approach is not CP-based, but, it depends on the CP rules for the critical conflict analysis.

11.1.3.1   Implications and BCP

Let C be a normalized LPB constraint

$$a_1 x_1 + \ldots a_n x_n \geq b. \tag{11.6}$$

That is, $b \geq a_1 \geq \ldots \geq a_n \geq 1$. Note that x_i's in C should be considered the literals of the variables in the original constraint before normalization. While literals may be assigned to 0 or 1 in decisions, they are assigned only to 1 in implications.

Two situations are of interest to BCP, the occurrences of conflicts and the implications. Let L represents the set of literals $\{x_1, \ldots, x_n\}$ of C in (11.6). Then C is unsatisfiable iff

$$\sum_{\substack{x_j \in L}}^{x_j \neq 0} a_j < b \tag{11.7}$$

and for an unassigned variable x_i, C implies $x_i = 1$ iff

$$\sum_{\substack{x_j \in L}}^{x_j \neq 0} a_j - a_i < b. \tag{11.8}$$

A consequence of the condition (11.8) is that, if x_i is implied, then any x_j such that $a_j > a_i$ is also implied. Therefore, one LPB constraint may produce multiple implications in one BCP, as opposed to at most one implication per clause per BCP in SAT. Let a_{max} be the largest coefficient where x_{max} is unassigned, then there is an implication in C iff x_{max} is implied by C, that is, iff

$$\sum_{\substack{x_j \in L}}^{x_j \neq 0} a_j - a_{max} < b. \tag{11.9}$$

This gives the weakest condition for any possible implication when all literals are considered. Furthermore, since this condition is also weaker than the conflict condition (11.7), it suffices to use just this condition to trigger checks for implications and conflicts.

For a constraint with many literals, implications and conflicts tend to occur only after many of its literals are set to false. If such "long" constraints are typical of a 01-ILP problem, the BCP process can be improved by "watching" only a subset of the literals for their propagation. The condition in (11.9) can be easily extended to implement this scheme. Let L' be a set of non-false (true or unassigned) literals in L, and a'_{max} the largest coefficient of an unassigned literal in L'. Then we can watch the literals in L' for the following condition

$$\sum_{\substack{j \in L'}}^{x_j \neq 0} a_j - a'_{max} < b. \tag{11.10}$$

It is safe to use the above condition because it is even weaker than the one in (11.9).[2] Figures 11.2 and 11.3 give an implementation of watched literals.

---

[2] To see this, consider the possibilities of a_{max}. It can be the coefficient of a literal in L' or in $L \setminus L'$. In the first case, a_{max} must equal a'_{max}; and since $L' \subseteq L$, it is clear that (11.9) implies (11.10). In the second case, the same implication follows from $L' \subseteq L \setminus \{x_{max}\}$.

```
// n:  number of literals in the constraint
// p->watch:  flag for watched literals
// p->value:  value of literals

watch(l) {
    if (update_watch(DELETE, l)) return UPDATED;
    for (p=1; p<=n; p++)
        if (!p->watch)
            p->watch = 1;
            if (update_watch(ADD, p)) return UPDATED;
    if (is_conflict()) return CONFLICT;

    for (p=1; p<=n; p++)
        if (p->watch)
            p->value = 1;
            if (update_watch(IMPLY, p)) return IMPLIED;
    return IMPLIED;
}
```

Figure 11.2.   Constraint watching for normalized LPB constraints.

The procedure *watch(l)* is invoked when the literal l is assigned to false. The basic idea is to try to update the watched set, with the procedure *update_watch*, so that the condition (11.10) is not triggered. In doing so, we end up with one of three results: the watched set is updated with no further actions, a conflict is encountered, or new assignments are implied.

11.1.3.2   Conflict Analysis

Conflict analysis in 01-ILP proceeds in almost the same way as does conflict analysis in CNF-based SAT, as shown in Chapter 5. They all work backwards on the implication graph to deduce a reason for the conflict. The only difference is that the former deduces using the CP rules whereas the latter uses resolution. This difference, however, incurs a non-trivial overhead in dealing with the "non-asserting" problem: deduction using the cutting-plane rules may lead to a learned constraint that does not flip any decision (see Section 5.6.3 of Chapter 5).

We have already seen an example where a logic conflict cannot be detected without using the rounding rule. Here is another example from [CK03] that demonstrates the "learning" of a non-asserting constraint. Consider an 01-ILP

```
// sum:LHS of the inequality (11.10)
// pmax:watched lit.with the largest coef.
// next_watch():  returns next watched lit.
//    a dummy lit.of coef.0 is returned
//    if there are no more literals.

is_conflict() {
    return sum + pmax->coef < d;
}

update_watch(type, p) {
    if (type == DELETE)
        p->value = 0;
        if (p != pmax)
            sum -= p->coef;
        else
            pmax = next_watch(p);
            sum -= pmax->coef;
    if (type == ADD)
        sum += p->coef;
        if (p->coef > pmax->coef)
            sum -= (p->coef - pmax->coef);
            pmax = p;
    if (type == IMPLY)
        if (p == pmax)
            pmax = next_watch(p);
            sum += (p->coef - pmax->coef);

    return sum < d;
}
```

Figure 11.3. Updating watched literals for normalized LPB constraints.

problem with three constraints:

$$
\begin{array}{rlr}
(a) & 3x_1 + 2x_2 + x_7 + 2\overline{x}_8 & \geq \ 3 \\
(b) & 3\overline{x}_1 + x_3 + x_5 + x_9 & \geq \ 3 \\
(c) & \overline{x}_2 + \overline{x}_3 + x_6 & \geq \ 2
\end{array}
$$

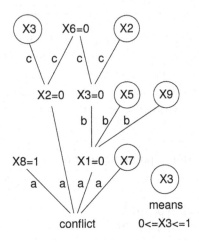

Figure 11.4.   Example: an implication graph.

Let $x_6 = 0$ be the current decision and $x_8 = 1$ an earlier decision. The constraint propagation leads to a conflict, at constraint (a), as shown in the implication graph in Figure 11.4. Following the reverse implication order x_1, x_2, and x_3, one wishes to eliminate the literals using the combination (a) + (b) + 2(c) + (c). The result would be

$$\overline{x}_3 + x_5 + 2x_6 + x_7 + 2\overline{x}_8 + x_9 \geq 4. \tag{11.11}$$

Note that x_3 is not removed, although it is one of the implied variables in the implication graph. This is an artifact of multiple implications from one constraint in ILP (the constraint (c) implies $x_2 = 0$ and $x_3 = 0$). This new constraint in (11.11) is non-asserting because it is satisfiable under the current assignments to x_3, x_6 and x_8: examination of the conflict condition (11.7) gives $4 < 4$, meaning there is no conflict. Thus no further information can be derived to guide the backtracking.

The above analysis, however, leads to a key observation: starting from the conflicting constraint, we also must arrive at a conflicting constraint to gain any additional information. To this end, we use the quantity

$$\sum_{x_j \in L}^{x_j \neq 0} a_j - b, \tag{11.12}$$

a rewriting of condition (11.7), to indicate whether or not a constraint is conflicting, and in a sense, by how much. This quantity is called the *slack* of the constraint in [CK03]. Therefore, a constraint is conflicting iff its slack is negative. We can now make more observations about slack and devise a strategy for obtaining asserting constraints.

1. All constraints except the conflicting constraint used in the implication graph have non-negative slacks (because BCP stops at the first conflict).

2. The cutting-planes of constraints with non-negative slacks also have non-negative slacks (otherwise the inference would be wrong).

It follows that the slack obtained in conflict analysis will stay non-negative once it becomes non-negative. Therefore, the slack must be kept negative at all times. We know this can always be achieved because the cutting-plane is a complete inference method. And, fortunately, this is achievable by simply repeating the following *reduction* and *saturation* steps to reduce the slack of a participating constraint.

Reduction: Pick a non-false literal x_i, and reduce its coefficient and the RHS accordingly. This is equivalent to adding $\lambda(-x_i \geq -1)$, where λ is the amount of the reduction. Consider $3x_1 + 2x_2 \geq 3$, where x_2 is non-false. Then the reduction on x_2 can give $3x_1 + x_2 \geq 2$ or $3x_1 \geq 1$. Note that the slack does not change, it stays as -1 if $x_1 = 0$ or +2 if $x_1 = 1$.

Saturation: The same as the *coefficient reduction* introduced in Section 11.1.1. Saturation simply replaces all coefficients larger than the RHS with the RHS. Saturation is equivalent to a sequence of rounding. Note that saturation reduces slack only if the coefficient of a non-false literal has been reduced. For example, $3x_1 + x_2 \geq 2$ is saturated to $2x_1 + x_2 \geq 2$. The slack is reduced from 1 to 0 if $x_1 = 1$, but stays as -1 if $x_1 = 0$.

The two steps basically apply the CP rules given in (11.3) and (11.5). The last example seems to suggest that the slack of certain constraints is non-reducible, which is problematic because maintaining a negative slack in conflict analysis requires the reducibility of positive slacks down to and including 0. The following lemma guarantees the satisfiability of this requirement.

LEMMA 11.1 In an implication graph, the slacks of all the constraints, except the conflicting constraint, can be reduced to 0.

PROOF: All constraints in an implication graph preceding the conflicting constraint are *inferring*, that is, they infer some variable assignment. Let C be such a constraint. A constraint of 0-slack can be obtained by simply reducing all the true and unassigned literals from C and performing saturation. To see this, let L_0 be the set of false literals and L_1 the set of true and unassigned literals excluding the implied literal x_k. Reducing all literals in L_1 gives

$$\sum_{x_i \in L_0} a_i x_i + a_k x_k \geq d - \sum_{x_i \in L_1} a_i.$$

Let $d - \sum_{x_i \in L_1} = d'$. We know $a_k \geq d' > 0$ because C is non-conflicting and inferring. After saturation we have

$$\sum_{x_i \in L_0} a_i' x_i + d' x_k \geq d'$$

which has a slack that equals 0.   □

The construction in the proof gives a simple approach to learning that guarantees an asserting constraint at the end. Intuitively, the result is the "exact" reason of the conflict because we have removed all the non-false literals which are not relevant to the implications and ultimately the conflict. However, literal reduction weakens the result. In the previous example, if we turn constraints (b) and (c) into 0-slack constraints as in the proof,[3] we eventually get the constraint

$$3x_6 + x_7 + 2\overline{x}_8 \geq 3. \tag{11.13}$$

Whereas a "larger" constraint

$$\overline{x}_3 + 3x_5 + 4x_6 + 2x_7 + 4\overline{x}_8 \geq 7 \tag{11.14}$$

is obtained if we only remove the literal x_9 from (b), which is just enough to keep the slack negative. It is interesting to note that the strength of the constraint (11.14) and that of the constraint (11.13) cannot be ordered. The former constrains two more variables, x_3 and x_5, and appears to be stronger. But the latter implies "if $x_6 = 0$ then $x_8 = 0$ and $x_7 = 1$", which is not implied by the former.

There are two more types of learning, both resulting in weaker constraints than the ones learned in the same manner as (11.14). The weakest learned constraint, and the easiest to learn, is one that corresponds directly to a cut in the implementation graph, just like in CNF learning. In this example, the result would be

$$x_6 + \overline{x}_8 \geq 1. \tag{11.15}$$

Another type of learned constraints are the so-called *cardinality constraints* [Bar95b] that can be inferred from any constraint (e.g., from the learned constraint (11.14)). A cardinality constraint is one in which at least k literals are true:

$$x_1 + x_2 + \ldots + x_n \geq k, \quad n \geq k. \tag{11.16}$$

Given an LPB constraint in the normal form (11.6), a cardinality constraint can be constructed by adding up the largest coefficients until the sum first surpasses

---

[3](c) generates two 0-slack constraints, one for $\overline{x}_2$ and one for $\overline{x}_3$, which eventually resolve both the literals. This is not guaranteed in other approaches. This is typical for UIPs in CNF-SAT (Section 5.6.3, Chapter 5, which contains at most one literal from the current decision level.

the RHS. The number of coefficients added is the cardinal k, and the literals are the ones of the original LPB constraint. For example, from constraint (11.14), one can infer the cardinality constraint

$$\overline{x}_3 + x_5 + x_6 + x_7 + \overline{x}_8 \geq 2. \tag{11.17}$$

Cardinality constraints tend to capture much of the information carried in the original constraint. From the above constraint, one can deduce that out of any four literals, one must be true. This corresponds to five four-literal clauses. However, in the context of conflict analysis, cardinality constraints may become non-asserting because of the weakening in their derivation. And this is what occurs in (11.17), which has a slack of 1. To avoid this, the same reduction-and-saturation procedure must be applied to the original constraint (11.14). In this specific example, the asserting cardinality constraint is equivalent to the one in (11.15) that is learned structurally from the implication graph.

The advantage of cardinality constraints is that the propagation and implications in them closely resemble those of the clauses. For example, a cardinality constraint with a cardinality k adopts a $(k+1)$-literal-watching scheme that is a straightforward extension of the 2-literal-watching scheme of clauses (Section 5.6.3, Chapter 5). In addition, like clauses, the cardinality constraints encountered in conflict analysis are not processed for slack reduction because their slacks are always zero.

11.2 Multi-valued Satisfiability

We will give a brief review of satisfiability involving multi-valued (MV) variables in [Hoo94, LK03]. The satisfiability problem, MV-SAT, is formulated similar to CNF-SAT except that the clauses, called MV clauses, are based on MV variables. An MV clause is defined as a disjunction of the form

$$x_1^{v_1} + \ldots + x_n^{v_n}, \tag{11.18}$$

where $x_1, \ldots, x_n$ are the MV variables, $x_i^{v_i}$ is a literal of x_i with respect to a set of integers v_i, whose logical valuation equals the membership test $x_i \in v_i$. Each variable x_i has a finite domain d_i. Without loss of generality, we assume that all variables share the same domain $\{0, \ldots, m-1\}$. Thus the literal $x_1^{\{2,4\}}$ is true iff x_1 is either 2 or 4.

The valuation of an MV literal $x_i^{v_i}$ extends to a set of variable assignments u_i as follows:

$$x_i^{v_i} \mid u_i = \begin{cases} True & \text{if } u_i \subseteq v_i \\ False & \text{if } u_i \cap v_i = \emptyset \\ Unknown & \text{otherwise} \end{cases} \tag{11.19}$$

Essentially, evaluation over a set captures the result of the evaluation over a value non-deterministically picked from the set.

Constraint propagation in MV-SAT employs the same unit-clause rule as in CNF-SAT: a literal is implied iff all of the other literals are false. Also, a conflict occurs iff all literals in a clause are false. Consequently, the watched-literals scheme in CNF-SAT can be directly used in MV-SAT.

The resolution of the two clauses $C_1 + x_k^{v_k}$ and $C_2 + x_k^{u_k}$ over variable x_l gives the clause

$$C_1 + C_2 + x_k^{v_k \cap u_k} \tag{11.20}$$

Conflict analysisconflict analysis is also done similarly to CNF-SAT, by resolving variables in the reverse implication order in the implication graph (IG). Because of the unit-clause implication rule and the conflict rule stated before, constraints in the IG contain no "Unknown" literals.

With all these similarities, there is a key difference in conflict analysis between a MV-SAT and a CNF-SAT solver. In the latter, a clause can be learned by simply collecting the literals on a cut in the IG, skipping resolutions as an optimization. In a MV-SAT solver, however, resolutions must be "played out" to avoid the weakening problem in learned clauses. Consider an example taken from [LK03]. To avoid lots of brackets, we write, e.g., $x_1^{\{2,4\}}$ as $x_1^{2,4}$. The variables in the following clauses have the domain $\{0, 1, 2, 3\}$.

$$
\begin{align}
(1) &\quad x_3^1 + x_2^{0,1} \\
(2) &\quad x_4^{1,3} + x_3^0 + x_1^1 \\
(3) &\quad x_4^3 + x_2^2 + x_1^{2,3}
\end{align}
$$

These clauses are in conflict under $x_4 = \{0\}$ and $x_3 = \{3\}$. Immediately, one can invert x_4 and x_3 and learn the clause $x_4^{1,2,3} + x_3^{0,1,2}$. However, resolving (3) and (2) on x_1, then resolving the result and (1) on x_2 give the clause $x_4^{1,3} + x_3^{0,1}$, which is tighter.

This phenomenon can be explained by viewing the MV-SAT as an ILP problem and by simulating resolutions with cutting-planes. It has been shown in the previous section that cutting-planes can carry literals (the non-false ones) that did not really cause the conflict, but nonetheless were learned in the final cutting-plane. The analogy in MV resolution is that some hidden variables used in converting MV clauses to ILP inequalities are added to make the learned clause tighter. A simple conversion enumerates an MV literal with a disjunct of Boolean variables. For example, $x_4^{1,3}$ becomes $x_{41} + x_{43}$. The clauses (2) and (3) from above are the equivalent to

$$
\begin{align}
(1') &\quad x_{31} + x_{20} + x_{21} &\geq&\ 1 \\
(2') &\quad x_{41} + x_{43} + x_{30} + x_{11} &\geq&\ 1 \\
(3') &\quad x_{43} + x_{22} + x_{12} + x_{13} &\geq&\ 1
\end{align}
$$

together with constraints representing the domain of the original variables. For x_1 the constraint, in negated form, would be

$$-x_{10} - x_{11} - x_{12} - x_{13} = -1.$$

To resolve x_1 from (2') and (3'), we add the two inequalities and the domain constraint of x_1, obtaining

$$x_{41} + 2x_{43} + x_{30} + x_{22} \geq 1 + x_{10} \geq 1$$

Similarly, resolving x_2 using this result and (1') yields

$$x_{41} + 2x_{43} + x_{30} + x_{31} \geq 1 + x_{23} \geq 1$$

After saturating the coefficient of x_{43}, we retrieve the learned clause in terms of the original variables $x_4^{1,3} + x_3^{0,1}$.

With the about enumeration, MV-SAT and 01-ILP should have similar complexity. The key to the efficiency of the former is thus to avoid enumeration as much as possible. For small domains, one may consider one-hot encoding for the value sets.

11.3 Word-level Constraint Solving

We have discussed logic-based constraint solving in 01-ILP and MV-SAT, both of which can represent certain problems more efficiently than CNF or the corresponding bit-level netlist. The logic-based approaches are closely related and are in fact the direct adoption of the DPLL-based CNF-SAT approach. In word-level constraint solving, while the same trend continues, there are other aspects that are of interest. First, word-level constraints encompass all constraints that are not bit-level. They must be converted to a form suitable for the underlying solver. Second, word-level and bit-level constraints usually coexist, representing the typical data and control partitioning. Of the three possible combinations, i.e., coercing both into the integer or the Boolean domain, or solving cooperatively in the constraints' native domain, this last, hybrid, approach characterizes most of the recent attempts in word-level constraint solving. The hybrid solvers in [FDK98, ABC+02, PICW04, BBC+05] can be captured in a very abstract way in Figure 11.5. As can be seen, this flow essentially adds to the standard DPLL flow a generic (integer) linear constraint solving capability. This capability is needed not only to check the feasibility of word-level constraints, but also to provide feedback to the conflict analysis engine. In addition, the Boolean propagation in DPLL becomes propagation in both the Boolean and word domains in the hybrid flow, which in turn requires modification to conflict analysis. And since there is a rich set of linear constraint methods, what methods to use and how to use them are also valid considerations in the hybrid solver. Finally, the circuit based reasoning and decision heuristics in ATPG are also helpful in word-level constraint solving.

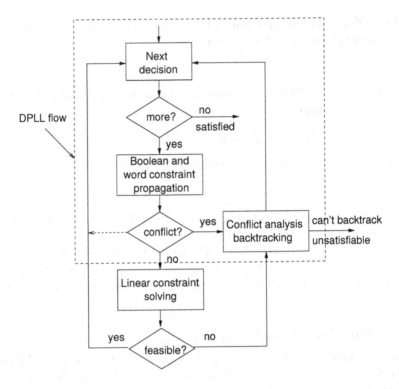

Figure 11.5.   Hybrid word-level constraint solving.

11.3.1   Converting RTL to Integer Linear Constraints

We are concerned with constraints that are expressed with word-level, RTL constructs. The first step is the synthesis of the RTL descriptions to a netlist in which the logical relation among signals is defined. A netlist can be seen as a DAG, where the directed arcs represent signals and their direction, and the nodes represent the operations on these signals. A signal can be single- or multi-bit, corresponding to a bit or a word. Accordingly, there are bit and word operations that apply to bits and words, respectively. Bit operations are the Boolean connectives such as AND, OR, XOR, and NOT. Word operations include comparing (including equality), addition, subtraction, multiplication, division, shifting, concatenation, and extraction (one or multiple bits of a word are used as a separate signal). Note that such bit-wise operations as the bit-wise OR and the MUX (multiplexer), are macro operations. The former, although applied to words, is actually a sequence of bit operations, whereas the latter is a composite of bit and word operations.

A word-level netlist preserves all of the above RTL operations. A bit-level netlist expands the word-level operations and thus contains only bit-level operations. In Chapter 5 we gave a simple example of the synthesis from RTL to a word- and bit-level netlist. We also showed in that chapter how ATPG directly works on a bit-level netlist, and how the netlist maps to a collection of clauses. To review this, we have listed the clauses for the AND, OR, NOT and XNOR gates in Table 11.1. Deriving clauses for the first three gates is straightforward. Note that the XNOR gate has only four clauses. This shows the effect of collapsing when clauses are derived from the XNOR function. In a straightforward "compositional" approach, an XNOR, which is equivalent to an AND of two OR gates, would have resulted in nine clauses.

Table 11.1. Clauses of the AND, OR, and NOT gates.

Gate	Definition	Clauses
AND	$a = b \cdot c$	$(\bar{a} + b)(\bar{a} + c)(a + \bar{b} + \bar{c})$
OR	$a = b + c$	$(a + \bar{b})(a + \bar{c})(\bar{a} + b + c)$
NOT	$a = \bar{b}$	$(a + \bar{b})(\bar{a} + b)$
XNOR	$a = (b + \bar{c})(\bar{b} + c)$	$(a + b + c)(a + \bar{b} + \bar{c})(\bar{a} + \bar{b} + c)(\bar{a} + b + \bar{c})$

Most of the word operations just mentioned can be captured directly in integer linear constraints. Others require certain remodeling, usually by introducing fresh variables. Also, all word operations use modular arithmetic. In the following, we assumed the words in a word operation would have the same width, but if not, we could left-pad the shorter word with 0s to make them of equal width. We denote a word operator with the corresponding integer operation with a subscript indicating the width of its result, e.g., $a +_n b$. This width can be arbitrarily specified. By default, however, it is equal to the width of the operands.

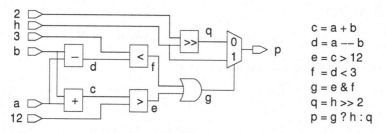

Figure 11.6. An example of a word-level netlist and corresponding constraints.

A word-level netlist corresponds to a conjunctive collection of constraints representing the word-level operations and Boolean gates. An example is given in Figure 11.6. The word-level constraints are then converted to integer linear constraints, as we will show in this section. The handled word operations are listed in Table 11.2. Others operations, such as division by variable, will have to be expanded and handled at the bit-level.

Table 11.2.   Word-level operations.

Operation	Description	Linearity
$a \% b$	Modulo, b is a constant;	Non-linear
$a +_n b$	Addition	Linear
$a -_n b$	Subtraction	Linear
$a *_n b$	Multiplication	Non-linear
$a/_n b$	Division by constant (floor(a/b))	Linear
$a >>_n, <<_n b$	left, right shifting with 0s; b is a constant.	Linear
$a ==_n, ! =_n b$	Equality, inequality.	Linear
$a >=_n, =<_n, >, < b$	Equality, inequality.	Linear
$\{a, b\}$	Concatenation	Non-linear
$a[i:j]$	Extraction. i and j are constants.	Non-linear
$a?b : c$	Multiplexing. a is Boolean.	Linear

11.3.1.1   A Generic Conversion with Modularization

The linear word-level constraints only need to be modularized, using the standard definition of the modulo operation. Let $\circ_n$ be an n-bit word operation and $\circ$ be the corresponding operation on integers. Then we have

$$a \circ_n b = (a \circ b) \bmod m, \text{where } m = 2^n. \tag{11.21}$$

For example, $2 + 3 = 5$. and $2 +_2 3 = 5 \bmod 4 = 1$. From the definition of modulo, we have

$$0 \leq a \circ_n b = (a \circ b) - mx \leq m - 1, \text{where } x \in \mathbb{Z}. \tag{11.22}$$

from which we derive a constraint about x:

$$\frac{a \circ b}{m} - (1 - \frac{1}{m}) \leq x \leq \frac{a \circ b}{m}. \tag{11.23}$$

Combining the equality (11.21) and the inequality (11.22), we can replace a linear word operation with the corresponding modulated integer operation and a constraint, as shown in the following conversion rule.

$$\frac{a \circ_n b}{a \circ b - mx \mid 0 \leq a \circ b - mx \leq m - 1}.$$

Depending on the range of $a \circ b$, an additional constraint about x can be deduced from the inequality (11.23). This constraint is not required, but it can help constraint solving.

11.3.1.2 Deriving Linear and Modular Constraints

First, each word variable is associated with a domain constraint defined by its declared width.

Word: Each n-bit word a becomes integer a with domain $[0, 2^n]$, that is

$$\frac{a, \ an \ n\text{-}bit \ word}{a \ | \ 0 \le a \le 2^n - 1}.$$

The conversions of the word operations in Table 11.2 are given in the following.

Modulo: The rule is the direct application of the definition of modulo.

$$\frac{a \% b}{a - bx \ | \ 0 \le a - bx \le b - 1}. \tag{11.24}$$

Addition: Consider $a +_n b$, where a and b are both n-bit words. Let $m = 2^n$. Then we have $0 \le a + b \le 2(m - 1)$. Applying this range to the inequality (11.23), we get $0 \le x \le 1$. Therefore, we have the addition conversion rule:

$$\frac{a +_n b}{a + b - mx \ | \ 0 \le a + b - mx \le m - 1, 0 \le x \le 1}.$$

Note that if we keep the carry bit, that is, let the result be $n + 1$-bit, then we expect $a +_n b$ to be equivalent to just $a + b$. This can be confirmed by changing m to $2m$ (for the $n + 1$-bit range) in the inequality (11.23), which would give $x = 0$. The same non-modular conversion, by adjusting the width of the result, also applies to other word operations. Similarly, m can be an arbitrary positive integer and the rule (11.3.1.1) still holds.

Subtraction: We rewrite $a -_n b$ as the addition of a and 2's complement of b, as $a -_n b + 2^n + 1$.[4] Then the addition conversion rule (11.3.1.2) is applied to $a -_n b$.

Multiplication with a constant: Consider $a *_n k$, k is an integer constant. Applying $0 \le ka \le km, m = 2^n$ to the inequality (11.23) gives $0 \le x \le k$. Therefore,

$$\frac{a *_n k}{ka - mx \ | \ 0 \le ka - mx \le m - 1, 0 \le x \le k - 1}.$$

---

[4]In case of underflow, the sign bit is ignored. Signed words can be handled by adding bit operations on the sign bits, parallel to the word operations.

Division with a constant: We introduce a fresh integer variable, y, to represent the result of $a/_n k$. Assume the integer division rounds down (the rounding-up case is similar). Then $a - ky = a \bmod k$, and we have

$$\frac{a/_n k}{y \mid 0 \leq a - ky \leq k - 1}.$$

Equalities, inequalities, and comparisons: We only need to treat the inequality and the comparisons ">" and "<". We can replace $a >_n b$ with $a >=_{n+1} b + 1$, where n is the width of a and b. The new width $n + 1$ is used to prevent possible overflow in $b+1$. This is valid since these operations return a Boolean, and thus the width is not dictated by the results. "<" is handled similarly. Conversion of the inequality involves the *additive inverse* [BD02] of word variables. The idea is that $a \mathrel{!=}_n b$ iff $a = b +_n x$ and $x \mathrel{!=} 0$ where x is an n-bit word. The modularity is handled by adding another variable y, as follows.

$$\frac{a \mathrel{!=}_n b}{a = b + x - 2^n y, \ 1 \leq x \leq 2^n - 1, \ 0 \leq y \leq 1}.$$

Shifting: By default, 0s are shifted in. In this case, a left $a <<_n k$ is the multiplication $a *_n k$, and a right shift $a >>_n k$ is the division $a/_n k$. Note that n takes the width of a by default. Then if instead 1s are shifted in (with operator $<<^1$, the result is the sum of the corresponding shift-in-0 result with the shifted-in 1s. Specifically, $a <<^1_n k = a <<_n k + 2^k$ and $a >>^1_n k = a >>_n k + 2^n - 2^{n-k}$.

Concatenation: Unlike other operations, the width of a concatenation is the sum of the width of its operands. Therefore, there is no over-flow and no modularization is needed. Consider the concatenation $\{a, b\}$, where b is an n-bit word. The conversion rule is simply

$$\frac{\{a, b\}}{2^n a + b}.$$

Extraction: Let a be an n-bit word. The extraction $a[i : j], i \geq j$ is the $(i - j + 1)$-bit word $(a <<_n (n - i - 1)) >>_{i-j+1} (n - i + j - 1)$, or equivalently, $(a *_n 2^{n-i-1})/_{i-j-1} 2(n - i + j - 1)$. Applying the multiplication rule, we have the replacement for $(a *_n 2^{n-i-1})$ as:

$$2^{n-i-1} a - 2^n x \mid 0 \leq 2^{n-i-1} a - 2^n x \leq 2^n - 1, \ 0 \leq x \leq 2^{n-i-1}.$$

Then applying the division rule on $2^{n-i-1} a - 2^n x/_{i-j+1} 2^{n-i+j-1}$, we get the extraction result

$$y \mid 0 \leq 2^{n-i-1} a - 2^n x - 2^{n-i+j-1} y \leq 2^{n-i+j-1} - 1.$$

Note that this constraint for y subsumes the previous one for x ($0 \leq 2^{n-i-1}a - 2^n x \leq 2^n - 1$) since $2^{n-i+j-1}y + 2^{n-i+j-1} - 1 \leq 2^n - 1$. Therefore, the conversion rule for extraction is

$$\frac{a[i:j]}{y \mid 0 \leq 2^{n-i-1}a - 2^n x - 2^{n-i+j-1}y \leq 2^{n-i+j-1} - 1, \; x \leq 2^{n-i-1}}.$$

Multiplexing: Let $Mux_n(s, a, b)$ mean "if s then the result equals a, else it equals b" where s is a bit and a and b are n-bit words. Let $m = 2^n$, the conversion rule is

$$\frac{Mux_n(s, a, b)}{y \mid \begin{array}{l} y - a - m(1 - x) \leq 0, \; a - y - m(1 - x) \leq 0, \\ y - b - mx \leq 0, \; b - y - mx \leq 0, \; 0 \leq x \leq 1 \end{array}}$$

Alternatively, we can apply a bit-wise AND of s to every bit of a, and of $\bar{s}$ to every bit of b, and add the two results as the multiplexing result. This way, multiplexing is converted linear constraints using the addition rule.

Multiplication with two variables: Finally, for a multiplication with two variables, we expand one of the operants so that the multiplication becomes a shift addition of partial products, as shown below.

$$\begin{aligned} a *_n b = \; & Mux(b_0, a, 0) +_n \\ & Mux(b_1, a <<_n 1, 0) +_n \cdots +_n \\ & Mux(b_{n-1}, a <<_n (n-1), 0). \end{aligned}$$

where b_i, $0 \leq i \leq n - 1$, is the ith bit of b.

11.3.2 Propagation and Implication

In the hybrid SAT flow discussed earlier, decisions are made in the control logic and propagated in both the control logic and data path. We will now discuss in more detail the propagation and implication in the data path, i.e., among integer linear constraints. We will consider integer linear constraints of the following normal form

$$\sum_{i=1}^{n} a_i x_i \geq b, \quad a_i, b, x_i \in \mathbb{Z}. \tag{11.25}$$

Integer variables, or word signals, from an RTL circuit or constraints have finite domains. We define a domain as an inclusive interval, specified by an upper and a lower bound.[5] Domains may be explicit or implicit. Implicit

---

[5] Assigning to a bit signal x, where x is an extraction from a word signal or a part of a concatenation that forms a word signal, effectively fragments the domain of the word signal. However, this fragmentation is made implicit by the modeling of extraction and concatenation that introduces fresh variables (Section 11.3.1).

bounds are determined by the sign and precision RTL signals, e.g., the domain of an unsigned 8-bit signal is $[0:255]$, and a bit signal has the domain $[0:1]$. An *empty* domain is one whose lower bound is greater than its upper bound.

An assignment to a finite domain variable reduces the domain by moving the bounds inward. A *complete* assignment reduces the domain to a singleton. Assignments are the results of branching decisions or implications from constraint propagation. A Boolean domain may be reduced at most once. Further reduction leads to an empty domain and causes a conflict. On the other hand, the domain of a word variable may be reduced multiple times. The domain of a variable shrinks monotonically between two consecutive backtrackings.

Assignments, i.e., bound changes, are propagated in the constraints so that further bound changes may be implied, or a conflict may be derived. Propagation in a constraint c is facilitated by two quantities, the maximum and minimum possible valuation of the LHS of c, denoted by $Max(c)$ and $Min(c)$, respectively. Let c be the inequality $\sum_{i=1}^{n} a_i x_i \geq b$, where x_i has the domain $[x_i^l, x_i^u]$. Then $Max(c)$ and $Min(c)$ are computed as follows.

$$Max(c) = \sum_{i, a_i > 0} a_i x_i^u - \sum_{i, a_i < 0} a_i x_i^l \qquad (11.26)$$

$$Min(c) = \sum_{i, a_i > 0} a_i x_i^l - \sum_{i, a_i < 0} a_i x_i^u \qquad (11.27)$$

A conflict occurs iff

$$Max(c) < b. \qquad (11.28)$$

The constraint is satisfied iff

$$Min(c) \geq b. \qquad (11.29)$$

An implication on x_i is a necessary domain change so that domain consistency is maintained. The domain of x_i is consistent with c iff for every value in the domain, there are values in the domains of all the other variables in c which constituent a solution to c. This consistency can be achieved by eliminating every value v in the domain of x_i such that the sum of v and the maximum valuation of the LHS without x_i is less than the RHS. Depending on the sign of a_i, this elimination process effectively increases the lower bound or decreases the upper bound of x_i.

Let us look at the case when $a_i > 0$. The maximum valuation of the LHS less x_i is $Max(c) - a_i x_i^u$. An implication is made on x_i if there is a v in its domain such that $Max(c) - a_i x_i^u + a_i v < b$. All such values can be removed if we increase the lower bound of x_i such that $Max(c) - a_i x_i^u + a_i x_i^l \geq b$. We assign the new lower bound $x_i^{\prime l}$ the minimum satisfying integer value so as to retain all possible values in the domain. Therefore, the implication is

$x_i'^l = \lceil x_i^u - \frac{Max(c)-b}{a_i} \rceil$. A similar result can be obtained for the case $a_i < 0$. The overall implication condition and results are given below.

$$Max(c) - |a_i| \, (x_i^u - x_i^l) < b \rightarrow \begin{cases} x_i'^l = \lceil x_i^u - \frac{Max(c)-b}{a_i} \rceil & \text{if } v < 0 \\ x_i'^u = \lfloor x_i^l - \frac{Max(c)-b}{a_i} \rfloor & \text{else} \end{cases}$$

(11.30)

Note that a conflict occurs iff some implication forces the lower bound to be higher than the upper bound, resulting in an empty domain. Therefore, one can opt to check implications and conflicts together by virtue of the above rule.

It should be noted that a solving procedure is complete only if it is capable of reducing the domains of all variables, bit- or word-level, to containing exactly one value. Thus a hybrid solver that makes decisions only in the Boolean domain must be completed by a linear constraint solver. The HDPLL solver in [PICW04] and [IPC05] uses the *Omega-test* [Pug92], which is based on FME (Section 11.1.2). The MATHSAT solver in [BBC+05] uses a series of LP and ILP solvers. Such solvers are used after the propagation just discussed finishes without finding a conflict. The solvers either find a final solution, or detect a conflict since the earlier propagation is not capable of detecting all conflicts.

11.3.3 Lazy Evaluation

Lazy evaluation is meant to minimize the amount of computation in propagation. Minimization can happen in two areas: the conflict and implication examination per constraint, and the selection of constraints that must be examined. Lazy evaluation in integer linear constraints is a generalization of the watched-literal method in the CNF-SAT and SAT-based 01-ILP discussion in Section 11.1.3.1.

From the implication rule (11.30), if x_i is implied, then any x_j such that $|a_j| \, (x_j^u - x_j^l) \geq |a_i| \, (x_i^u - x_i^l)$ is also implied. Let x_m be the variable such that $|a_m| \, (x_m^u - x_m^l)$ is the maximum among all the variables in constraint c. Therefore, there is an implication iff there is an implication on x_m, that is, iff

$$Max(c) - |a_m| \, (x_m^u - x_m^l) < b.$$

(11.31)

The implication checks on other variables are carried out only if this condition is satisfied. Furthermore, since a conflict occurs iff some implication forces an empty domain, this condition also suffices for triggering the detection for conflicts.

Propagation can be further improved by examining only a subset of constraints. This is done by selecting a subset of "watched" variables from each constraint such that the constraint is checked only when some of these variables are assigned. Let c' be the constraint obtained from c by keeping only the watched variables,[6] and a variable x'_m in c' defined the same way as x_m in c. Then we can watch the variables in c' for the following condition

$$Max(c') - |a'_m| \, (x'^u_m - x'^l_m) < b. \tag{11.32}$$

Using this condition, a procedure similar to the watched-literal procedure for 01-ILP (See Figure 11.2) maintains the watched set and detection implications and conflicts. It should be noted that the watched-set is useful only in problems where long constraints are typical. In addition, the incurred overhead may offset the gain even in these cases.

11.3.4   Multiple Domain Reductions

Propagation of a decision may reduce the domain of a word signal multiple times, and, in the worst case, as many times as there are values in the original domain. The following example shown in Figure 11.7 demonstrates such an extreme case. The initial decision sets the output of the comparator ($\geq$) to 1. Due to the chain of implications from $w_2 = w_3 + 2$, $w_1 = w_3 + 1$, and $w_1 \geq w_2$, the domain of w_3 is iteratively tightened, each time by the amount of 1. Reaching the empty domain takes approximately 2^n iterations, where n is the width of w_3. However, with FME or the cutting-plane rules (Section 11.1.2), the conflict can be derived in a straightforward way, as in the following ($x = y$ is replaced with $x \geq y$ and $y \geq x$).

$$
\begin{array}{rcl}
w_3 + 1 & \geq & w_1 \\
w_1 & \geq & w_2 \\
w_2 & \geq & w_3 + 2 \\
\hline
1 & \geq & 2
\end{array}
\tag{11.33}
$$

It is therefore necessary to limit the depth of propagation and resort to an inference procedure such as FME, or *negative-cycle detection* [CG99], to avoid enumerating domains. Note that a generic ILP solving must be applied after constraint propagation anyway.

11.3.5   Conflict Analysis

Conflict analysis in a hybrid word-level constraint solver has the following scenarios:

1. The conflict occurs during propagation in the bit-level logic only.

---

[6]An optimization is to exclude all fixed variables (whose domain contains exactly one value) from the watched set because we do not branch on them and any implication of a different assignment would have caused a conflict immediately.

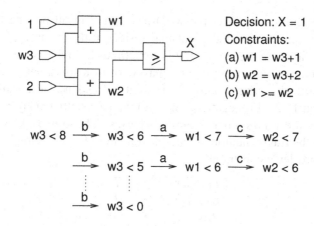

Figure 11.7. An example of multiple domain reductions.

2. The conflict occurs during propagation that has already affected the word-level logic.

3. The conflict occurs in the generic ILP solver, which provides an implication/resolution graph for the conflict.

4. The conflict occurs in the generic ILP solver, which provides a minimal set of conflicting constraints.

5. The conflict occurs in the generic ILP solver, which provides no further information about the conflict.

Conflict analysis in case 1 is identical to that of CNF-SAT. In cases 2 and 3, ideally, an analysis based on slack computation similar to the one in Section 11.1.3.2 can be applied. However, the multiple domain reduction problem (Section 11.3.4) complicates this computation and can cause considerable overhead. Therefore, cases 2 to 5 are usually handled using structural analysis, i.e., by finding a small cut in the implication/resolution graph of the conflict.

An example for the generic ILP solver in case 3 is the Omega-test [Pug92] based-on FME. Both MATHSAT [BBC+05] and HDPLL [IPC05] use the Omega-test. Note that, however, an infeasible ILP problem may satisfy the FME check and only fail the integral requirement (Section 11.1.2). If this happens, case 3 degenerates to case 5. In case 4, conflict analysis relies on the ILP solver to produce a minimal conflicting set. For example, MATHSAT also uses an LP solver based on the Cassowary algorithm [BMSX97] (a variant of the simplex method), which efficiently computes the minimal conflicting set. In the last case, conflict analysis starts from all the bit signals at the bit/word boundary.

The propagation may affect the word-level logic in all but the first case. Although one can always find a cut that resides fully in the bit-level logic (since all decisions are made there), it is sometimes desirable to cut across the bit/word boundary. For example, one may wish to learn about word signals used for control or configuration. Such a cut corresponds to a multi-valued clause (Section 11.2). Propagation, implication, and conflict analysis in multi-valued clauses are very similar to normal clauses. The drawback, as we have mentioned, is the fragmentation of variable domains. For an example, consider the follow word-level constraints.

$$
\begin{array}{rcl}
2x_1 + 5x_2 & \geq & 7 \\
-5x_2 + x_3 & \geq & 3 \\
-2x_1 - x_3 & \geq & -11
\end{array}
\tag{11.34}
$$

Suppose all variables have the same integer domain $[0, 15]$. Suppose a propagation in other constraints not shown leads to the implications $x_1 \leq 3$ and $x_1 \geq 2$. They cause a conflict in the three constraints above. The multi-valued clause corresponding to a cut at x_1 is $x_1^{\{[0,1],[4,15]\}}$.

11.3.6   Arithmetic Solving

So far we have been focusing on SAT techniques in word-level constraint solving such as propagation, implication, and conflict analysis. However, these techniques alone do not constitute a complete decision procedure - they must be helped by a generic ILP or arithmetic solver. Since such a solver can be invoked as frequently as the SAT solver (see Figure 11.5), its efficiency is also as important. In addition, the arithmetic solver's capability of diagnosing the conflict source is also critical to conflict analysis on the SAT side.

Fortunately, there is wide range of choices for arithmetic solvers. For example, HDPLL uses an ILP solver based on FME which provides a resolution graph for a conflict encountered during variable elimination. Both ICS and CVC Lite use variants of Shostak's decision procedure for quantifier-free first-order logic that can be viewed as a generalization of Gaussian elimination. CVC Lite also provides an "abstract proof" that shows a small set of constraints responsible for a conflict.

MATHSAT has the most comprehensive arithmetic solving capability among the works surveyed. It uses a six-stage arithmetic solver, as shown in Figure 11.8 (from [BBC+05]).

The six stages are divided into three groups that work in the domains of uninterpreted functions, linear arithmetic over reals, and linear arithmetic over integers.

- Equality over Uninterpreted Functions (EUFs)

- Linear Arithmetic over Reals

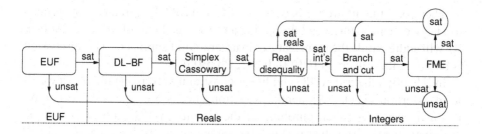

Figure 11.8. The MATHSAT arithmetic solver flow.

- Bellman-Ford search for negative cycles [CG99] in Difference Logic (DL-BF): generates a small conflict set

- Simplex (using the Cassowary algorithm [BMSX97]): generates a minimal conflicting set

- Disequality

• Linear Arithmetic over Integers

- Branch-and-cut

- FME

This arrangement not only accommodates mixed LP and ILP solving, but also "filters" the problem so that less general but faster procedures are invoked first. In addition, reasons learned from conflicts earlier are used in later stages to prune the solution space.

11.4 ATPG-based Word-level Constraint Solving

ATPG and SAT are both search-based solutions to propositional constraints, as we have shown in Chapter 5.[7] Many techniques from ATPG or SAT are readily adaptable by the other. For example, the recursive learning [KP94] originated in ATPG can also be used as a preprocessing in SAT (e.g., in [AS00]), and conflict analysis from SAT may also be implemented for ATPG. One strength of ATPG algorithms is their awareness of structural information. However, once SAT solvers have access to such information, they also perform well in traditional test generation problems [TGH97, GF01] and in generic constraint solving for circuits [GAG+02].

---

[7]We consider only the constraint solving aspect of ATPG corresponding to fault sensitization (Section 5.7, Chapter 5). Fault propagation, or the observability problem, is not a concern here.

Therefore, the difference between ATPG and SAT for circuits is superficial to a large extent, so instead of discussing a full-fledged word-level ATPG solver, we will address only the main features in its structural approach. This discussion is based on the solvers in [HC00, Iye03, PICB05].

In [HC00], a hybrid ATPG and linear constraint solver is applied to a word-level assertion check. Sequential assertion checks are converted to combinational ones by implicit unrolling of the circuit. ATPG cooperates with an arithmetic constraint solver in much the same way as the hybrid SAT and ILP flow discussed earlier. Starting with the assertion as the first unjustified gate, the solver iterates through the following steps:

1. Traverse backwards and find a *cut*, i.e., a set of decision points at the primary inputs, flip-flop outputs, high-fanout nets, and control/data interface signals such as comparator outputs and multiplexer selects.

2. Make decisions and propagate to justify all unjustified gates. Backtrack the last decision upon reaching a conflict. Declare the overall problem as unsatisfiable if all decisions have been backtracked. Otherwise, call the arithmetic solver if all unjustified gates are justified.

3. The arithmetic solver tries to justify the data path. If a solution is found, then the decision points, excluding the primary inputs, become the new set of unjustified gates, and the process resumes from Step 1. If no such decision points exist, declare the overall problem as satisfiable.

The above process is captured in Figure 11.9.

Since ATPG works directly off a netlist, propagation and implication are performed as forward and backward 3-valued logic simulation. Essentially, simulation models for all bit- and word-level operations must be developed, with some care, especially in the backward implication cases.

The word-level constraint solver in [Iye03], specifically implemented for random vector generation, is unique in that it does not resort to a generic arithmetic constraint solver. The whole solving process is done within the branch-and-bound framework of ATPG. The word-level constraints are first converted to a circuit in the straightforward way, after which additional sign and precision handling logics are inserted. Propagation takes place in both the control logic and the data path. The propagation is pessimistic, for various efficiency reasons. For example, it may just give up in the presence of multiple-domain-reductions (Section 11.3.4) that last too long, or simply return the complete range of a word signal due to algorithmic difficulties or too many fragmented intervals in the range. The fall-back for this pessimism is the guarantee that all single-valued propagations are exact. Note that for completeness in the absence of a generic arithmetic solver, decisions must be made on both the control and data-path variables.

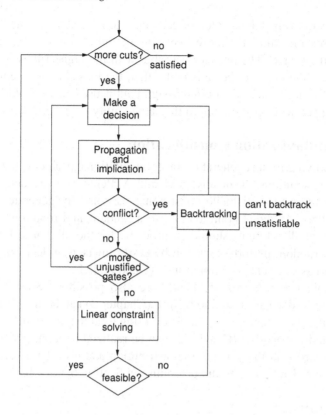

Figure 11.9. A word-level hybrid ATPG-ILP flow.

In addition to propagation, decision making is another key ingredient in ATPG that utilizes structural information. A common strategy in ATPG is that decisions are made only on key variables, as opposed to possibly on all variables in SAT.[8] The usual heuristic for decision variables is to select primary inputs and high fanout nets. In word-level constraint solving, such interface signals as multiplexer selects and comparator outputs are also good candidates.

From a pool of candidates, the solver in [HC00] uses biases to select the decision variable and a value appropriate to the nature of the problem: whether or not the problem is more likely to be satisfiable. The bias for each decision variable is computed recursively starting from the gate to be justified. Suppose we are justifying a 1 at the output of an OR gate. Then the probability for one of its inputs to be 1 is 2/3 because the input is 1 in 2 out of the 3 input assignments that make the output 1. A similar computation can be carried out for other types

---

[8]That is, for SAT solvers that have not adopted this strategy!

of gates. Recursively, a bias value of each decision variable is computed such that the higher the value, the more likely the biased assignment (towards 1 or 0) justifies the target. The decision variable with the highest bias is selected. If one expects the target to be satisfiable, then the biased assignment is used. Otherwise, the negation of the biased assignment is used. Both intend to prune the search space with speculation on the satisfiability of the problem.

11.5   Randomization Consideration

Constraint solving for test generation faces an additional problem: the randomization of the solutions. Typically, SAT and ATPG solvers are deterministic. That is, for the same satisfiable constraints, they usually generate the same solutions. This is a direct contradiction to a fundamental requirement in test generation that all solutions should be probable for the sake of test coverage. Fortunately, random perturbations can be easily injected in decision makings (variable and assignment selections) in SAT and ATPG.

In addition, useful forms of biasing are also possible. Specifically, the variable-solve-order based randomization (Chapter 3) can be implemented by simply forcing the order on the variable ordering used in the decisions. In addition, weighted distribution (Chapter 3) can be realized by adding new random variables to represent the items in the distribution and give these variables the highest solve order, similar to the procedure in Section 6.5 of Chapter 6.

11.6   Summary

We have given an in-depth survey of recent word-level constraint solving approaches in the EDA field. Common to these approaches is the adoption of the recent developments in SAT, mainly the ideas of efficient propagation and implication, and conflict analysis. In addition, structural decision heuristics from ATPG are also shown to add to the effectiveness. There are other important techniques, such as efficient problem formulation, constraint implication, and static learning, which we omitted but feel warrant further study.

Typical word-level constraint solving deals with a hybrid environment, signified by the usual coexistence of control and data-path logics in EDA applications. While many of the surveyed works thoroughly investigate the SAT-based approach which lends itself well to the bit-level control logic, more effort seem to be in order on the arithmetic solving side. The MATHSAT solver [ABC$^+$02, BBC$^+$05], with its hierarchy of solvers handling multiple theories in the real and integer domains, is evidence of the effort that has been made in this direction.

Appendix A
Acronyms

Table A.1. Acronyms

Acronyms	Words
01-ILP	Integer Linear Programming over Booleans
ABV	Assertion-Based Verification
ADD	Algebraic Decision Diagram
AP	Atomic Proposition
ATPG	Automatic Test Pattern Generation
BCP	Boolean Constraint Propagation
BDD	Binary Decision Diagrams
BFS	Breadth-First Search
BMD	Binary Moment Diagram
BNF	Backus Naur Form
BU	Boolean Unification
CBV	Cycle-based-Verilog
CLP	Constraint Logic Programming
CNF	Conjunctive Normal Formal
CP	Cutting-plane
CSP	Constraint Satisfaction Problem
CTL	Computational Tree Logic
DAG	Directed Acyclic Graph
DFA	Deterministic Finite Automaton
DFS	Depth-First Search
DLIS	Dynamic Largest Individual Sum
DPLL	Davis-Putnam-Logemann-Loveland algorithm

(*cont.*)

Table A.2.   Acronyms (*cont.*)

Acronyms	Words
DUV	Design Under Verification
EDA	Electronic Design Automation
ESL	Electronic System Level
EUF	Equality over Uninterpreted Function
FAN	Fanout-Oriented algorithm
FL	Foundation Language (of PSL)
FME	Fourier-Motzkin Elimination
FSM	Finite State Machine
GC	Generalized Cofactoring
GDL	General Description Language
HDD	Hybrid Decision Diagrams
HVL	High Level Verification
iff	if and only if
IG	Implication Graph
ILP	Integer Linear Programming
IPC	Irredundant Prime Cover
lfp	least fixed point
LPB	Linear Pseudo Boolean
LP	Linear Programming
LRM	Language Reference Manual
LTL	Linear Time Logic
MTDD	Multi-Terminal Decision Diagram
MV-SAT	Multi-Valued Satisfiability
NFA	Nondeterministic Finite Automaton
OBE	Branching Extension (of PSL)
OOP	Object-Oriented Programming
OVL	Open Verification Library
PLI	Program Language Interface
PNF	Positive Normal Form
PODEM	Path-Oriented Decision Making
PSL	Property Specification Language
RAM	Random Access Memory
RE	Regular Expression
ROBDD	Reduced Ordered Binary Decision Diagram
RT	Register Transfer
RTL	Register Transfer Level
SAT	Propositional Satisfiability
SCV	SystemC Verification library
SERE	Extended Regular Expression
SSA	Single Stuck-At fault
SVA	SystemVerilog Assertion
SVRC	SystemVerilog Random Constraint
TD	Tree-decomposition
UIP	Unique Implication Point
UF	Uninterpreted Function
VSIDS	Variable State Independent Decaying Sum

Appendix B
Proofs

B.1 Proofs for Chapter 6

LEMMA 6.5   Given a constraint BDD f and a state s, $\omega(r(f), s) = \sum_{\alpha \in f_s} \pi(\alpha, s)$ where f_s is the set of legal vectors in f under the s.

PROOF: We prove by induction on the number of variables in f. First, let $sum(f, s)$ abbreviate $\sum_{\alpha \in f_s} \pi(\alpha, s)$.

If f is a function with no variables, thus a constant, then for any s: (1) if $f = 1$, then $\omega(ONE, s) = 1$, also, $sum(1, s) = 1$ since all input vectors are legal; (2) similarly, if $f = 0$, both $\omega(ZERO, s)$ and $sum(0, s)$ are 0.

If f has one variable, there are also two cases: (1) if the variable is an input, without loss of generality, let $f = x$. Then $\omega(x, s) = p^x(s)$, which is equal to $sum(x, s)$ since $x = 1$ is the only legal vector; (2) if the variable is a state variable, without loss of generality, let $f = y$. Then $\omega(y, y = 1)$ returns the weight of ONE, 1, and $sum(y, y = 1)$ is also 1 since all vectors are legal; similarly, $\omega(y, y = 0)$ is the weight of $ZERO$, 0, and $sum(y, y = 0)$ is 0 since no vectors are legal.

Now, we prove the induction hypothesis. Let the lemma hold for the two child BDDs of f, g, and h, of variables $u_1, \ldots, u_{n-1}$; let u_n be the new variable in f, and, without loss of generality, let $f = u_n \cdot g + \overline{u_n} \cdot h$. Again we have two cases.

(1) u_n is an input variable: from Equation (6.2), we have

$$\omega(r(f), s) = p^{u_n}(s) \cdot \omega(r(g), s) + p^{\overline{u_n}}(s) \cdot \omega(r(h), s).$$

By the induction hypothesis, we get

$$\omega(r(f), s) = p^{u_n}(s) \cdot sum(g, s) + p^{\overline{u_n}}(s) \cdot sum(h, s)$$

The right-hand-side, due to Definition 6.2, becomes

$$p^{u_n}(s) \cdot \sum_{\alpha \in g_s} \prod_{i=1}^{n-1} \left[\alpha_i \cdot p^{u_i}(s) + (1-\alpha_i) \cdot p^{\overline{u_i}}(s) \right] +$$
$$p^{\overline{u_n}}(s) \cdot \sum_{\alpha \in h_s} \prod_{i=1}^{n-1} \left[\alpha_i \cdot p^{u_i}(s) + (1-\alpha_i) \cdot p^{\overline{u_i}}(s) \right]$$

$$= \quad \sum_{\beta \in u_n \cdot g_s} \prod_{i=1}^{n} \left[\beta_i \cdot p^{u_i}(s) + (1-\beta_i) \cdot p^{\overline{u_i}}(s) \right] + \quad ;; \beta_n = 1$$
$$\sum_{\alpha \in \overline{u_n} \cdot h_s} \prod_{i=1}^{n} \left[\alpha_i \cdot p^{u_i}(s) + (1-\alpha_i) \cdot p^{\overline{u_i}}(s) \right] \quad ;; \beta_n = 0$$

$$= \quad \sum_{\alpha \in f_s} \prod_{i=1}^{n} \left[\alpha_i \cdot p^{u_i}(s) + (1-\alpha_i) \cdot p^{\overline{u_i}}(s) \right]$$

$$= \quad \sum_{\alpha \in f_s} \pi(\alpha, s)$$

$$= \quad sum(f, s)$$

(2) u_n is a state variable: If $s' = s \cdot u_n$, then $f_{s'} = g_s$, and therefore $sum(f, s') = sum(g, s)$; furthermore, from Equation (6.2), we have $\omega(f, s') = \omega(g, s)$. Hence, by the induction hypothesis, $\omega(f, s') = sum(f, s')$. The case $s' = s \cdot \overline{u_n}$ can be proved similarly. $\square$

THEOREM 6.8   Under a legal state, the procedure *Walk* only visits nodes with positive weights.

PROOF: We use an inductive argument. First, from Lemma 6.6, under any legal state the weight of the root is greater than 0; furthermore, by Definition 6.4, any node with a positive weight must have at least one child node with a positive weight. Therefore, due to Definition 6.7, at least one of its branching probabilities is greater than 0; so the traversal in *Walk* from that node must take a branch with a positive (not a 0) possibility, thus reaching a node with a positive weight. By induction, every visited node must have a positive weight. $\square$

THEOREM 6.10   The procedure *Walk* generates input vectors according to their constrained probabilities.

PROOF: First, from Corollary 6.9, *Walk* always ends at node *ONE*. Therefore, it never generates illegal input vectors, whose constrained probabilities are 0 according to Definition 6.3.

   Now, consider legal vectors. Let f be the constraint and s the state. Let $\sigma_1, \ldots, \sigma_m, \sigma_{m+1}$ be the sequence of nodes visited in a *Walk* traversal, where σ_1 is the root and σ_{m+1} the node *ONE*. Without loss of generality, assume that the input variables corresponding to this sequence are $x_1, \ldots, x_m$, and that $x_{m+1}, \ldots, x_n$ were not visited. Additionally, let $\alpha_1 \cdots \alpha_n$ be the generated vector, wherein the first m values correspond to the branches taken in the traversal, and the last $n - m$ values correspond to the choices based upon input probabilities. For brevity, let $p^{u,b,s}$ denote $b \cdot p^u(s) + (1 - b) \cdot p^{\overline{u}}(s)$. Then, the probability of this vector is given in the following product: $\frac{p^{x_1, \alpha_1, s} \cdot \omega(\sigma_2, s)}{\omega(\sigma_1, s)} \cdot \frac{p^{x_2, \alpha_2, s} \cdot \omega(\sigma_3, s)}{\omega(\sigma_2, s)} \cdots \frac{p^{x_m, \alpha_m, s} \cdot \omega(\sigma_m, s)}{\omega(ONE, s)} \cdot \prod_{j=m+1}^{n} p^{x_j, \alpha_j, s}$. This product simplifies to $\left(\prod_{i=i}^{n} p^{x_i, \alpha_i, s} \right) / \omega(\sigma_1, s)$. Due to Definition 6.2 and Lemma 6.6, the last expression rewrites to $\pi(\alpha, s) / \sum_{\beta \in f_s} \pi(\beta, s)$. By Definition 6.3, this is exactly the branching probability of the legal input vector α. $\square$

LEMMA 6.12   Using the p-tree algorithm, the probability of generating an input vector in which an input variable x_k equals 1 (resp. 0) monotonically increases as p^{x_k} (resp. $p^{\overline{x_k}}$) increases.

PROOF: There are two cases: (1) if x_k is not visited in *Walk*, then it is assigned according to its input probabilities, to which the probability of the vector is proportional, resulting in the lemma being true; (2) if x_k is visited in *Walk*, and if f is the constraint and s

is the state, then the probability of the vector, say $\alpha = \alpha_1 \cdots \alpha_n$ where α_k is the value of x_k, is $\frac{\prod_{i=0}^{n}(\alpha_i \cdot p^{x_i}(s) + (1-\alpha_i) \cdot p^{\overline{x_i}}(s))}{\omega(r(f),s)}$. Because this probability does not depend on variable ordering (Corollary 6.11), we choose x_k to be the variable associated with the root node. The probability becomes $\frac{\prod_{i=1}^{n}(\alpha_i \cdot p^{x_i}(s) + (1-\alpha_i) \cdot p^{\overline{x_i}}(s))}{(p^{x_k}(s) \cdot \omega(t(r(f)),s) + p^{\overline{x_i}}(s) \cdot \omega(e(r(f)),s))}$. Let q denote the product of $(\alpha_i \cdot p^{x_i}(s) + (1-\alpha_i) \cdot p^{\overline{x_i}}(s))$ for all $i \neq k$, w_t denote $\omega(t(r(f)),s)$, and w_e denote $\omega(t(r(f)),s)$. Note that q, w_t, and w_e are independent of the input probabilities of x_k. Letting $\alpha_k = 1$, the above formula rewrites to $(q \cdot p^{x_k}(s)) / (p^{x_k}(s) \cdot w_t + p^{\overline{x_k}}(s) \cdot w_e)$ $q / (w_t + (p^{\overline{x_k}}(s)/p^{x_k}(s)) \cdot w_e)$ $q / (w_t + (1/p^{x_k}(s) - 1) \cdot w_e)$. Therefore, the probability of α monotonically increases in p^{x_k}. The case $\alpha_k = 0$ is analogous. $\qquad\square$

THEOREM 6.13 Let C be a set of constraints, $C_1, \ldots, C_n$ be the disjoint-input-support partition of C, and $\alpha^1, \ldots, \alpha^n$ be the corresponding partial vectors generated by applying *p-tree* to the $C_1, \ldots, C_n$. Let α denote the concatenation $\alpha^1.\alpha^2 \ldots \alpha^n$. Then, under any state, the probability of generating $\alpha^1, \alpha^2, \ldots,$ and α^n is equal to the constrained probability of generating α from C.

PROOF: Let s be the state. Let f be the conjunction of all constraints, and $f^1, \ldots, f^n$ the conjunctions of constraints in the groups. Hence, $f = \bigwedge_{i=1}^{n} f^i$. First, we prove

$$\prod_{i=1}^{n} \sum_{\beta \in f_s^i} \pi(\beta,s) = \sum_{\beta \in f_s} \pi(\beta,s). \tag{B.1}$$

We know

$$\sum_{\beta \in f_s^1} \pi(\beta,s) \cdot \sum_{\gamma \in f_s^2} \pi(\gamma,s) = \sum_{\beta \in f_s^1} [\pi(\beta,s) \cdot \sum_{\gamma \in f_s^2} \pi(\gamma,s)]$$

$$= \sum_{\beta \in f_s^1} \sum_{\gamma \in f_s^2} [\pi(\beta,s) \cdot \pi(\gamma,s)]$$

$$= \sum_{\beta \in f_s^1} \sum_{\gamma \in f_s^2} \pi(\beta \cdot \gamma,s)$$

since f^i and f^j, where $i \neq j$, have disjoint input supports. Therefore, $\beta \in f_s^i \wedge \gamma \in f_s^j$ iff $\beta.\gamma \in f_s^i \wedge f_s^j$, the above formulas can be further reduced to

$$\sum_{\beta.\gamma \in (f_s^i \wedge f_s^j)} \pi(\beta \cdot \gamma,s) = \sum_{\delta \in (f^i \wedge f^j)_s} \pi(\delta,s).$$

We have proven Equation (B.1) for the case $n = 2$. Since $f^1 \wedge f^2$ and f^i for $i > 2$ again have disjoint input supports, by induction, Equation (B.1) holds for all n.

Equation (B.1) implies that $\sum_{\beta \in f_s} \pi(\beta, s) = 0$ iff there exists i, such that $\sum_{\beta \in f_s^i} \pi(\beta, s) = 0$. Therefore, an illegal state in f is also an illegal state in the partition and vice versa. So, in both cases, for the same set of illegal states, the probabilities for all (illegal) vectors are 0.

Now consider legal vectors. From Definition 6.2, we have $\prod_{i=1}^{n} \pi(\alpha^i, s) = \pi(\alpha, s)$. So when s is a legal state, dividing each side of the above equation by the corresponding side of Equation (B.1) gives: $\frac{\prod_{i=1}^{n} \pi(\alpha^i,s)}{\prod_{i=1}^{n} \sum_{\beta \in f_s^i} \pi(\beta,s)} = \frac{\pi(\alpha,s)}{\sum_{\beta \in f_s} \pi(\beta,s)}$ which is $\prod_{i=1}^{n} \frac{\pi(\alpha^i,s)}{\sum_{\beta \in f_s^i} \pi(\beta,s)} = \frac{\pi(\alpha,s)}{\sum_{\beta \in f_s} \pi(\beta,s)}$, where the left-hand side is the product of the probabilities of generating $\alpha^1, \ldots, \alpha_n$ from $C_1, \ldots, C_n$, respectively, and the right-hand side is the probability of generating α from C. Hence, the theorem also holds for legal vectors. $\qquad\square$

B.2   Proofs for Chapter 7

THEOREM 7.6   The conjunction of a set of prime det-constraints on x with mutually exclusive conditions $s_1, \ldots, s_l$ is a det-constraint, and vice versa. That is,

$$\bigwedge_{i=1}^{l} (s_i \rightarrow (x = b_i)) \leftrightarrow k \rightarrow (x = g) \tag{B.2}$$

where $s_i \wedge s_j = 0$ for $i \neq j$, $b_i \in \{0, 1\}$; k and g can be derived from the prime det-constraints as

$$k = \bigvee_{i=1}^{l} s_i, \; g \in [g^{on}, \overline{g^{off}}], \; g^{on} = \bigvee_{i=1}^{l} (b_i \cdot s_i), \; g^{off} = \bigvee_{i=1}^{l} (\overline{b_i} \cdot s_i), \tag{B.3}$$

and the prime det-constraints can be derived from k and g as in the set

$$\{(s_i, b_i) \mid 1 \leq i \leq |k|, \; s_i \in k, \; b_i = g_{s_i}\}. \tag{B.4}$$

where $|k|$ is the number of minterms in the onset of k.

PROOF: We prove the theorem by showing that, for both derivations, the valuations of the two sides of the equivalence in (B.2) under any state are equal.

Denote the state by s_j. If $s_j \notin k$, the theorem holds since both sides of (B.2) evaluate to 1 trivially, for both derivations. Conversely, if $s_j \in k$, we have, for both derivations,

$$\left(\bigwedge_{i=1}^{l} (s_i \rightarrow (x = b_i)) \right)_{s_j} = (x = b_j), \tag{B.5}$$

$$(k \rightarrow (x = g))_{s_j} = (x = g_{s_j}). \tag{B.6}$$

Now for the derivation in (B.3), if $b_j = 1$, then $s_j \in g^{on}$ and $g_{s_i} = 1$. Therefore both (B.5) and (B.6) evaluate to $x = 1$. Similarly, if $b_j = 0$, then $s_j \in g^{off}$ and $g_{s_i} = 0$, hence both (B.5) and (B.6) evaluate to $x = 0$.

For the derivation in (B.4), if $g_{s_j} = 1$, then $b_j = 1$, therefore both (B.5) and (B.6) evaluate to $x = 1$; similarly, if $g_{s_j} = 0$, then $b_j = 0$, therefore both (B.5) and (B.6) evaluate to $x = 0$.   □

THEOREM 7.9   For any det-constraint, $e := k \rightarrow (x = g)$, and Boolean function f, an input variable $y \neq x$ is extractable from $\tau(f, e)$ iff it is extractable from $f \cdot e$, or more precisely,

$$\frac{\partial (\exists_{y'} \tau(f, e))}{\partial y} = \frac{\partial (\exists_{y'} (f \cdot e))}{\partial y} \tag{B.7}$$

where $y' = X - \{y\}$.

PROOF: We only need to prove $\exists_{y'}(\tau(f,e)) = \exists_{y'}(f \cdot e)$. This equality is shown in the following computation. Note $f \cdot e = \tau(f,e) \cdot e$ because of Property 7.1.

$$
\begin{aligned}
\exists_{y'}(f \cdot e) &= \exists_{y'}(\tau(f,e) \cdot e) \\
&= \exists_{y'}((k \cdot f_{x=g} + \overline{k} \cdot f) \cdot (\overline{k} + (x = g))) \\
&= \exists_{y'}(\overline{k} \cdot f + k \cdot f_{x=g} \cdot (x = g) + \overline{k} \cdot f \cdot (x = g)) \\
&= \exists_{y'}(\overline{k} \cdot f + k \cdot f_{x=g} \cdot (x = g)) \\
&= \exists_{y'-\{x\}}(\exists_x(\overline{k} \cdot f + k \cdot f_{x=g} \cdot (x = g))) \\
&= \exists_{y'-\{x\}}(\exists_x(\overline{k} \cdot f) + \exists_x(k \cdot f_{x=g} \cdot (x = g))) \\
&= \exists_{y'-\{x\}}(\exists_x(\overline{k} \cdot f) + k \cdot f_{x=g} \cdot \exists_x(x = g)) \\
&= \exists_{y'-\{x\}}(\exists_x(\overline{k} \cdot f + k \cdot f_{x=g})) \\
&= \exists_{y'}\tau(f,e)
\end{aligned}
$$

$\square$

PROPERTY 7.1 $\tau(f,e) \cdot e = f \cdot e$, i.e., $\tau(f,e)$ is a function equivalent to f in the care set e.

PROOF: We need to prove

$$e \cdot \tau(f,e) = f \cdot e \tag{B.8}$$

First we compute the two sides of the above equation. The left-hand side is

$$
\begin{aligned}
e \cdot \tau(f,e) &= (k \to (x = g)) \cdot (k \cdot f_{x=g} + \overline{k} \cdot f) \\
&= (\overline{k} + (x = g)) \cdot (k \cdot f_{x=g} + \overline{k} \cdot f) \\
&= \overline{k} \cdot f + k \cdot f_{x=g} \cdot (x = g).
\end{aligned}
$$

while the right-hand side is

$$
\begin{aligned}
f \cdot e &= f \cdot (k \to (x = g)) \\
&= f \cdot (\overline{k} + (x = g)) \\
&= f \cdot (\overline{k} + k \cdot (x = g)) \\
&= \overline{k} \cdot f + k \cdot f \cdot (x = g).
\end{aligned}
$$

Then, for Equation (B.8) to hold, we only need to prove

$$f_{x=g} \cdot (x = g) = f \cdot (x = g) \tag{B.9}$$

which can be done again by expanding the two sides. Since the left-hand side is

$$
\begin{aligned}
f_{x=g} \cdot (x = g) &= (x \cdot f_x + \overline{x} \cdot f_{\overline{x}})_{x=g} \cdot (x = g) \\
&= (g \cdot f_x + \overline{g} \cdot f_{\overline{x}}) \cdot (x \cdot g + \overline{x} \cdot \overline{g}) \\
&= x \cdot g \cdot f_x + \overline{x} \cdot \overline{g} \cdot f_{\overline{x}},
\end{aligned}
$$

and the right-hand side is

$$
\begin{aligned}
f \cdot (x = g) &= (x \cdot f_x + \overline{x} \cdot f_{\overline{x}}) \cdot (x = g) \\
&= (x \cdot f_x + \overline{x} \cdot f_{\overline{x}}) \cdot (x \cdot g + \overline{x} \cdot \overline{g}) \\
&= x \cdot g \cdot f_x + \overline{x} \cdot \overline{g} \cdot f_{\overline{x}},
\end{aligned}
$$

therefore, Equation (B.9) holds, and consequently Equation (B.8) holds.   □

PROPERTY 7.2 $\tau(f, e)$ decreases the "diversity" of x in f, i.e., $\partial f/\partial x$, by the amount k. This implies that $\tau(f, e)$ is independent of x iff $k \geq \partial f/\partial x$

PROOF: We need to prove

$$\frac{\partial \tau(f, e)}{\partial x} = \overline{k} \cdot \frac{\partial f}{\partial x}. \tag{B.10}$$

First, we prove two lemmas: let g and h be two constraints, where g depends on x and h does not, then

(LEMMA 1) $\frac{\partial(g+h)}{\partial x} = \overline{g} \cdot \frac{\partial h}{\partial x}$

(LEMMA 2) $\frac{\partial(g \cdot h)}{\partial x} = g \cdot \frac{\partial h}{\partial x}$

The lemmas are proven as follows:

$$
\begin{aligned}
\frac{\partial(g+h)}{\partial x} &= (g+h)_x \cdot \overline{(g+h)_{\overline{x}}} + \overline{(g+h)_x} \cdot (g+h)_{\overline{x}} \\
&= (g + h_x) \cdot \overline{(g + h_{\overline{x}})} + \overline{(g + h_x)} \cdot (g + h_{\overline{x}}) \\
&= (g + h_x) \cdot \overline{g} \cdot \overline{h_{\overline{x}}} + \overline{g} \cdot \overline{h_x} \cdot (g + h_{\overline{x}}) \\
&= \overline{g} \cdot h_x \cdot \overline{h_{\overline{x}}} + \overline{g} \cdot \overline{h_x} \cdot h_{\overline{x}} \\
&= \overline{g} \cdot \frac{\partial h}{\partial x}
\end{aligned}
$$

and

$$
\begin{aligned}
\frac{\partial(g \cdot h)}{\partial x} &= (g \cdot h)_x \cdot \overline{(g \cdot h)_{\overline{x}}} + \overline{(g \cdot h)_x} \cdot (g \cdot h)_{\overline{x}} \\
&= (g \cdot h_x) \cdot \overline{(g \cdot h_{\overline{x}})} + \overline{(g \cdot h_x)} \cdot (g \cdot h_{\overline{x}}) \\
&= (g \cdot h_x) \cdot (\overline{g} + \overline{h_{\overline{x}}}) + (\overline{g} + \overline{h_x}) \cdot g \cdot h_{\overline{x}} \\
&= g \cdot h_x \cdot \overline{h_{\overline{x}}} + g \cdot \overline{h_x} \cdot h_{\overline{x}} \\
&= g \cdot \frac{\partial h}{\partial x}.
\end{aligned}
$$

Now, since

$$\tau(f, e) = k \cdot f_{x=g} + \overline{k} \cdot f$$

where $k \cdot f_{x=g}$ and $\overline{k}$ are independent of x, by applying the above lemmas, we have

$$
\begin{aligned}
\frac{\partial \tau(f, e)}{\partial x} &= \overline{k \cdot f_{x=g}} \cdot \overline{k} \cdot \frac{\partial f}{\partial x} \\
&= (\overline{k} + \overline{f_{x=g}}) \cdot \overline{k} \cdot \frac{\partial f}{\partial x} \\
&= \overline{k} \cdot \frac{\partial f}{\partial x}
\end{aligned}
$$

therefore, the property is true.   □

PROPERTY 7.3   If $\tau(f, e)$ is independent of x and $f \to e$, then $\tau(f, e) = \exists_x f$.

PROOF: We prove that if $\tau(f, e)$ is independent of x and $f \to e$, then $\tau(f, e) = \exists_x f$.

Since $f \rightarrow e$, i.e., $f \leq \overline{k} + (x = g)$, by cofactoring both sides with respect to x we get $f_x \leq \overline{k} + g$, which implies $k \cdot f_x \leq k \cdot g$, thus $k \cdot g \cdot f_x = (k \cdot g) \cdot (k \cdot f_x) = k \cdot f_x$. Similarly, $k \cdot \overline{g} \cdot f_{\overline{x}} = k \cdot f_{\overline{x}}$. Therefore, we have

$$
\begin{aligned}
k \cdot f_{x=g} &= k \cdot (g \cdot f_x + \overline{g} \cdot f_{\overline{x}}) & \text{(B.11)} \\
&= k \cdot (f_x + f_{\overline{x}}) & \text{(B.12)} \\
&= k \cdot \exists_x f. & \text{(B.13)}
\end{aligned}
$$

Furthermore, since $\tau(f, e) = k \cdot f_{x=g} + \overline{k} \cdot f$ is independent of x, by Property 7.2, we have

$$
\frac{\partial \tau(f, e)}{\partial x} = \overline{k} \cdot \frac{\partial f}{\partial x} = 0.
$$

Also, from the lemmas obtained in the proof of Property 7.2, we have

$$
\frac{\partial \overline{k} \cdot f}{\partial x} = \overline{k} \cdot \frac{\partial f}{\partial x}.
$$

Therefore, $\frac{\partial \overline{k} \cdot f}{\partial x} = 0$, thus $\overline{k} \cdot f$ is independent of x too. From this and the result in (B.13), we have

$$
\begin{aligned}
\tau(f, e) &= k \cdot f_{x=g} + \overline{k} \cdot f \\
&= k \cdot \exists_x f + \overline{k} \cdot f \\
&= \exists_x (k \cdot f) + \exists_x (\overline{k} \cdot f) \\
&= \exists_x f
\end{aligned}
$$

Hence, the property is true. $\qquad\square$

PROPERTY 7.4 If there exists a care set optimization function $\psi(f, e)$ that does not depend on x, then it must be $\tau(f, e)$.

PROOF: We prove that for any function $\psi(f, e)$ such that $\psi(f, e) \cdot e = f \cdot e$ and $\psi(f, e)$ is independent of x, then $\psi(f, e)$ is $\tau(f, e)$.

Let α be a minterm of variables in f and e. If $\alpha \in e$, then $(\psi(f, e) \cdot e)(\alpha) = (f \cdot e)(\alpha)$, i.e., $\psi(f, e)(\alpha) = f(\alpha)$. Similarly, since $\tau(f, e) \cdot e = f \cdot e$ (Property 7.1), we have $\tau(f, e)(\alpha) = f(\alpha)$. So ψ and τ agree in e.

Now we check the case $\alpha \in \overline{e}$, i.e., $\alpha \in (k \cdot (x \neq g))$. Let $\alpha' \in (k \cdot (x = g))$ be a minterm which differs from α only at variable x. From the definition of $\tau(f, e)$, we know $\tau(f, e)(\alpha) = f(\alpha')$. On the other side, for $\psi(f, e)$ to not depend on x, we must have $\partial(\psi)/\partial(x) = 0$, i.e., $\psi_x = \psi_{\overline{x}}$. Therefore, $\psi(f, e)(\alpha) = \psi(\alpha')$ (actually for all α). Now since $\alpha' \in (k \cdot (x = g)) \subseteq e$, we have $\psi(\alpha') = f(\alpha')$, thus $\psi(f, e)(\alpha) = f(\alpha')$. Hence, ψ and τ also agree in $\overline{e}$. $\qquad\square$

B.3   Proofs for Chapter 8

THEOREM 8.3 Under a legal state, any input assignment satisfying f^i is a suffix of an input assignment satisfying f^j, for all $j < i$.

PROOF: It suffices to prove for the case $j = i - 1$. Let α be an input assignment to $x_i, \ldots, x_n$ and β a legal state, such that $f^i(\alpha, \beta) = 1$. From the definition of *projections* of f, we have

$$f^i(x_i, \ldots, x_n, Y) = f^{i-1}(0, x_i, \ldots, x_n, Y) + f^{i-1}(1, u_i, \ldots, u_n, Y),$$

plugging in α and β, we have

$$f^i(\alpha, \beta) = f^{i-1}(0, \alpha, \beta) + f^{i-1}(1, \alpha, \beta).$$

The left-hand side of the above equation is 1, so at least one of the terms on the right-hand side must be true. That is, either $(0, \alpha)$ or $(1, \alpha)$, or both, under the state β, is a solution to f^{i-1}. $\square$

THEOREM 8.4   The solutions σ_i's connected as above defines a mapping $\sigma : \mathbb{B}^n \mapsto \mathbb{B}^n$, such that for any legal state and any pair of input vectors $\alpha, \alpha' \in \mathbb{B}^n$ and $\sigma(\alpha) = \alpha'$

1 α' satisfies f, and

2 if α satisfies f, then $\alpha' = \alpha$.

PROOF: We prove the first statement by induction on the length of α'. Let $\alpha' = \alpha'_1 \ldots \alpha'_n$. Since we are under a legal state, from the solution of x_i in Equation (9.9), we know that α'_n satisfies f^n. Assume we already have a partial solution $\alpha'_{i+1} \ldots \alpha'_n$ that satisfies f^{i+1}. By applying this solution to the solution of x_i, as indicated in the algorithm, and from Theorem 9.3, the resulting $\alpha'_i \ldots \alpha'n$ must satisfy f^i. Eventually we get an α' that satisfies f^1, and therefore f.

The second statement is proven by examining the solutions. Given a partial solution, α'_i has exactly one choice from u_i, 1, or 0. In the first choice, α'_i is equal to u_i no matter what value u_i takes; in the second choice, $\alpha'_i = 1$, and 1 is the only possible value for the ith position of an α that satisfies f. Therefore α'_i also takes 1. Similarly, both α'_i and α are 0 in the third choice. $\square$

THEOREM 8.9   The substitution returned by the algorithm in Figure 9.2 maps a vector α to a vector α' in the constraint that has the shortest distance (as given in Formula (9.15) from α.

PROOF: Suppose we have arrived at a partial solution $\alpha'_{i+1} \ldots \alpha'_n$, which is at a distance l from $\alpha_{i+1} \ldots \alpha_n$. According to the algorithm, the choice of α'_i either maintains the distance, or increases it to $l + 2^{i-1}$. If we were to do things differently and still produce a satisfying partial solution, we could (1) assign $\overline{u}_i$ to α'_i, or (2) backtrack and change an $\alpha'_j, j > i$, so that α'_i would not be forced to 1 or 0. However, (1) would increase instead of maintaining l, and (2) would increase l by 2^{j-1}, which is greater than the sum of any potential savings at positions $i < j$. $\square$

References

[ABC+97]   A. Aggoun, F. Bueno, M. Carro, P. Deransart, M. Fabris, W. Drabent, G. Ferrand, M. Hermenegildo, C. Lai, J. Lloyd, J. Maluszynski, G. Puebla, and A. Tessier. CP Debugging Needs and Tools. In *International Workshop on Automated Debugging*, pages 103–122, 1997.

[ABC+02]   G. Audemard, P. Bertoli, A. Cimatti, A. Kornilowicz, and R. Sebastiani. A SAT Based Approach for Solving Formulas over Boolean and Linear Mathematical Propositions. In *In Proceedings of the 18th International Conference on Automated Deduction (CADE-18), part of Federated Logic Conference (FLoC'02)*. Springer-Verlag, 2002.

[ABDD+90]   A. Aharon, A. Bar-David, B. Dorfman, E. Gofman, M. Leibowitz, and V. Schwartzburd. RTPG-A Dynamic Biased Pseudo-Random Test Program Generator for Processor Verification. In *IBM Technical Report 88.290*, July 1990.

[ABDD+91]   A. Aharon, A. Bar-David, B. Dorfman, E. Gofman, M. Leibowitz, and V. Schwartzburd. Verification of the IBM RISC System/6000 by a Dynamic Biased Pseudo-random Test Program Generator. In *IBM Systems Journal*, volume 30(4), pages 527–538, July 1991.

[ABG+00]   Y. Abarbanel, I. Beer, L. Gluhovsky, S. Keidar, and Y. Wolfsthal. FoCs: Automatic Generation of Simulation Checkers from Formal Specifications. In *Proceedings of the Computer Aided Verification Conference (CAV)*, volume 1855, pages 538–542, 2000.

[Acc03a]   Accellera. *Open Verification Library Assertion Monitor Reference Manual*. June 2003.

[Acc03b]   Accellera. *SystemVerilog 3.1 Language Reference Manual*. 2003.

[Acc04]   Accellera. *PSL 1.1 LRM*. June 2004.

[ACKS02]   G. Audemard, A. Cimatti, A. Kornilowicz, and R. Sebastiani. Bounded Model Checking for Timed Systems. In *In Proceedings of the 22nd Joint International Conference on Formal Techniques for Networked and Distributed Systems (FORTE 2002)*. Springer-Verlag, November 2002.

[Ake78]   S. B. Akers. Binary Decision Diagrams. In *IEEE Transactions on Computers*, volume C-37, pages 509–516, June 1978.

[ARMS02]   F.A. Aloul, A. Ramani, I.L. Markov, and K.A. Sakallah. Generic ILP versus Specialized 0-1 ILP: An Update. In *Proceedings of the Design Automation Conference (DAC)*, pages 450–457, June 2002.

[AS00]   F.A. Aloul and K.A. Sakallah. An Experimental Evaluation of Conflict Diagnosis and Recursive Learning in Boolean Satisfiability . In *Proceedings of International Workshop on Logic Synthesis*, 2000.

[ASSB94]   A. Aziz, V. Singhal, G. M. Swamy, and R. K. Brayton. Minimizing Interacting Finite State Machines: A Compositional Approach to the Language Containment Problem. In *Proceedings of International Conference on Computer Design (ICCD)*, pages 255–261, October 1994.

[ATB94]   A. Aziz, S. Tasiran, and R. K. Brayton. BDD Variable Ordering for Interacting Finite State Machines. In *Proceedings of the Design Automation Conference (DAC)*, June 1994.

[Bar95a]   P. Barth. A Davis-Putnam based Enumeration Algorithm for Linear Pseudo-Boolean Optimization. In *Technical Report MPI-I-95-2-003, Max-Planck-Institut Für Informatik*, 1995.

[Bar95b]   P. Barth. *Logic-Based 0-1 Constraint Solving in Constraint Logic Programming*. Kluwer, 1995.

[BB95]   P. Barth and A. Bockmayr. Finite domain and cutting plane techniques in CLP(PB). In L. Sterling, editor, *Proceedings of ICLP'95*, pages 133–147. MIT Press, 1995.

[BB04]   C. Barrett and S. Berezin. CVC Lite: A New Implementation of the Cooperating Validity Checker. In *Proceedings of the Computer Aided Verification Conference (CAV)*, 2004.

[BBC+05]   M. Bozzano, R. Bruttomesso, A. Cimatti, T. Junttila, P.V. Rossum, S. Schulz, and R. Sebastiani. Mathsat: Tight Integration of SAT and Mathematical Decision Procedures. In *Journal of Automated Reasoning, Special Issue on SAT*. Springer-Verlag, 2005.

[BC95]   R.E. Bryant and Y.-A. Chen. Verification of Arithmetic Circuits with Binary Moment Diagrams. In *Proceedings of the Design Automation Conference (DAC)*, pages 534–541, June 1995.

[BCMD92]   J. R. Burch, E. M. Clarke, K. L. McMillan, and D. L. Dill. Symbolic Model Checking: 10^{20} States and Beyond. In *Information and Computation*, volume 98(2), pages 142–170, 1992.

[BCW80]   M. Blum, A. Chandra, and M. Wegman. Equivalence of Free Boolean Graphs Can Be Decided Probabilistically in Polynomial Time. In *Information Processing Letters*, pages 10:80–82, 1980.

[BD97]   V. Bertacco and M. Damiani. The Disjunctive Decomposition of Logic Func-
            tions. In *Proceedings of International Conference on Computer-Aided Design
            (ICCAD)*, pages 78–82, 1997.

[BD02]   R. Brinkmann and R. Drechsler. RTL-Datapath Verification using Integer Lin-
            ear Programming. In *ASP-DAC '02: Proceedings of the 2002 conference on
            Asia South Pacific design automation/VLSI Design*, 2002.

[BDS91]   D. Bochmann, F. Dresig, and B. Steinbach. A New Decomposition Method
            for Multilevel Circuit Design. In *Proceedings of European Design Automation
            Conference (EURO-DAC)*, pages 374–377, 1991.

[BF89]   S. Bose and A. Fisher. Verifying Pipelined Hardware using Symbolic Logic
            Simulation. In *Proceedings of International Conference on Computer Design
            (ICCD)*, pages 217–221, 1989.

[BFG⁺93a]   R. I. Bahar, E. A. Frohm, C. M. Gaona, G. D. Hachtel, E. Macii, A. Pardo,
            and F. Somenzi. Algebraic Decision Diagrams and their Applications . In
            Proceedings of International Conference on Computer-Aided Design (ICCAD),
            pages 188–192, 1993.

[BFG⁺93b]   R.I. Bahar, E.A. Frohm, C.M. Gaona, G.D. Hachtel, E. Marcii, A. Pardo, and
            F. Somenzi. Algebraic Decision Diagrams and their Applications. In *Proceed-
            ings of International Conference on Computer-Aided Design (ICCAD)*, pages
            188–191, November 1993.

[BFH05]   D. Bustan, D. Fisman, and J. Havlicek. Automata Construction for PSL. Techni-
            cal report, The Weizmann Institute of Science, 2005. Technical Report MCS05-
            04.

[BFMY83]   C. Beeri, R. Fagin, D. Maier, and M. Yannakakis. On the Desirability of Acyclic
            Database Schemes. In *J. ACM*, volume **30**(3), pages 479–513, 1983.

[BGCZ99]   A. Biere, A. Gimatti, E. Clarke, and Y. Zhu. Symbolic Model Checking without
            BDDs. In *Proceedings of Tools for Algorithms for Construction and Analysis
            of Systems (TACAS)*, pages 193–207. Springer-Verlag, March 1999.

[BHSV⁺96]   R. K. Brayton, G. D. Hachtel, A. Sangiovanni-Vincentelli, F. Somenzi, A. Aziz,
            S.-T. Cheng, S. Edwards, S. Khatri, Y. Kukimoto, A. Pardo, S. Qadeer, R. K.
            Ranjan, S. Sarwary, T. R. Shiple, G. Swamy, and T. Villa. VIS: A system for
            Verification and Synthesis. In *Proceedings of the Computer Aided Verification
            Conference (CAV)*, July 1996.

[Bla38]   A. Blake. *Canonical Expressions in Boolean Algebra*. PhD thesis, University
            of Chicago, 1938.

[Blu05]   Bluespec. 2005. http://www.bluespec.com.

[BLW95]   B. Bollig, M. Löbbing, and I. Wegener. Simulated annealing to improve variable
            orderings for OBDDs. In *Proceedings of International Workshop on Logic
            Synthesis*, pages 5b:5.1–5.10, 1995.

[BM88]   R. S. Boyer and J. S. Moore. *A Computational Logic Handbook*. Academic
            Press, New York, 1988.

[BMD02]   T. Bengtsson, A. Martinelli, and E. Dubrova. A Fast Heuristic Algorithm for Disjoint Decomposition of Boolean Functions. In *Proceedings of International Workshop on Logic Synthesis*, pages 51–55, 2002.

[BMS00]   L. Baptista and J.P. Marques-Silva. Using Randomization and Learning to Solve Hard Real-world Instances of Satisfiability. In *Proceedings of the 6th International Conference on Principles and Practice of Constraint Programming (CP 00)*, September 2000.

[BMSX97]   A. Borning, K. Marriott, P. Stuckey, and Y. Xiao. Solving Linear Arithmetic Constraints for User Interface Applications. In *Proceedings of UIST 97*, pages 87–96, 1997.

[BRKM91]   K. M. Butler, D. E. Ross, R. Kapur, and M. R. Mercer. Heuristics to Compute Variable Ordering for the Efficient Manipulation of Binary Decision Diagrams. In *Proceedings of the Design Automation Conference (DAC)*, pages 417–420, June 1991.

[Bry86]   R. Bryant. Graph-based Algorithms for Boolean Function Manipulation. In *IEEE Transactions on Computers*, volume C-35, pages 677–691, August 1986.

[Bry91a]   R. E. Bryant. Formal Verification of Memory Circuits by Switch Level Simulation. In *IEEE Transactions on Computer-Aided Design of Integrated Circuits and Systems*, volume 10, No. 1, pages 94–102, January 1991.

[Bry91b]   R.E. Bryant. On the Complexity of VLSI Implementations and Graph Representation of Boolean Functions with Applications to Integer Multiplication. In *IEEE Transactions on Computers*, volume 40(2), pages 205–213, February 1991.

[But89]   M. Butts. 10, 100 and 1000 speed-ups in computers: What constraints change in CAD? In *Proceedings of International Conference on Computer-Aided Design (ICCAD)*, 1989. Presented at the Panel Session 2.

[Büc60]   J.R. Büchi. On a Decision Method in Restricted Second Order Arithmetic In *International Congress on Logic, Methodology, and Philosophy of Science*, pages 1–11, 1960.

[CAD]   CADENCE. *Verilog-XL User Reference.*

[CBM89]   O. Coudert, C. Berthet, and J. C. Madre. Verification of Sequential Machines Using Functional Boolean Vectors. In *Proceedings of the IFIP International Workshop, Applied Formal Methods for Correct VLSI Design*, November 1989.

[CE81]   E. M. Clarke and E. A. Emerson. Design and Synthesis of Synchronization Skeletons Using Branching Time Logic. In *Proceedings of Workshop on Logic of Programs*, volume 131 of *Lecture Notes in Computer Science*, pages 52–71. Springer-Verlag, 1981.

[CFZ95]   E.M. Clarke, M. Fijita, and X. Zhao. Hybrid Decision Diagrams - Overcoming the Limitations of MTBDDs and BMDs. In *Proceedings of International Conference on Computer-Aided Design (ICCAD)*, pages 159–163, 1995.

[CG99]   B. Cherkassky and A. Goldberg. Negative-Cycle Detection Algorithms. In *Mathematical Programming*, pages 277–311, 1999.

[CGZ96]   E.M. Clarke, S.M. German, and X. Zhao. Verifying the SRT Division Algorithm Using Theorem Proving Techniques. In *Proceedings of the Computer Aided Verification Conference (CAV)*, pages 111–122. Springer-Verlag, 1996.

[Che93]   K.-C. Chen. Boolean Matching Based on Boolean Unification. In *Proceedings of European Conference on Design Automation (EDAC)*, pages 346–351, 1993.

[CI92]   A. K. Chandra and V. S. Iyengar. Constraint Solving for Test Case Generation - A Technique for High Level Design Verification. In *Proceedings of International Conference on Computer Design (ICCD)*, pages 245–248, 1992.

[CJP83]   H. Crowder, E.L. Johnson, and M. Padberg. Solving Large-scale Zero-one Liner Programming Problems. In *Operations Research*, volume 31(5), pages 803–834, 1983.

[CK03]   D. Chai and A. Kuehlmann. A fast pseudo-boolean constraint solver. In *Proceedings of the Design Automation Conference (DAC)*, pages 830–835, 2003.

[CM90]   O. Coudert and J. C. Madre. A Unified Framework for the Formal Verification of Sequential Circuits. In *Proceedings of International Conference on Computer-Aided Design (ICCAD)*, pages 126–129, November 1990.

[CM92]   O. Coudert and J.C. Madre. Implicit and Incremental Computation of Primes and Essential Primes of Boolean Functions. In *Proceedings of the Design Automation Conference (DAC)*, pages 36–39, 1992.

[CM04]   K. Claessen and J. Martensson. An Operational Semantics for Weak PSL. In *Proceedings of the Formal Methods in CAD Conference (FMCAD)*, pages 337–351. Wiley, 2004.

[CMFT93]   O. Coudert, J.C. Madre, H. Fraisse, and H. Touati. Implicit Prime Cover Computation: An Overview. In *Proceedings of Synthesis and Simulation Meeting and International Interchange (SASIMI)*, 1993.

[CMZ+93]   E.M. Clarke, K. McMillan, X. Zhao, M. Fujita, and J. Yang. Spectral Transforms for Large Boolean Functions with Applications to Technology Mapping. In *Proceedings of the Design Automation Conference (DAC)*, pages 54–60, June 1993.

[Coo71]   S.A. Cook. The Complexity of Theorem Proving Procedures. In *Third Annual ACM SIGACT Symposium on the Theory of Computing*, pages 151–158, 1971.

[CT87]   W. Cook C.R. Coullard and G. Turan. On the complexity of cutting plane proofs. In *Discrete Applied Mathematics*, volume 18, pages 25–38, 1987.

[Dav84]   R. Davis. Diagnostic reasoning based on structure and behavior. In *Artificial Intelligence*, volume 24, pages 347–410, 1984.

[DBG95]   R. Drechsler, B. Becker, and G′ockel. A genetic algorithm for variable ordering of OBDDs. In *Proceedings of International Workshop on Logic Synthesis*, pages 5c:5.55–5.64, 1995.

[DBR96]   R. Drechsler, B. Beker, and S. Ruppertz. K*BMDs: A New Data Structure for Verification. In *The European Design and Test Conference*, spring 1996.

[DEF+04]   S. Dudani, C. Eisner, D. Fisman, H. Foster, J. Havlicek, J. Lu, E. Marschner, J. Märtensson, A. Sarkar, and B. Tabbara. Accellera FVTC Alignment Sub-Committee Final Report. In *Accellera*, February 2004.

[DGV99]   M. Daniele, F. Giunchiglia, and M.Y. Vardi. Improved Automata Generation for Linear Temporal Logic. In N. Halbwachs and D. Peled, editors, *Proceedings of the Computer Aided Verification Conference (CAV)*, volume 1633 of *Lecture Notes in Computer Science*, pages 249–260. Springer-Verlag, December 1999.

[DLL62]   M. Davis, G. Logemann, and D. Loveland. A Machine Program for Theorem-Proving. In *Communucations of ACM*, volume 5(7), pages 394–397, 1962.

[dMRS02]   L. de Moura, H. Rue, and M. Sorea. Lazy Theorem Proving for Bounded Model Checking over Infinite Domains. In *Proceedings of CADE*, 2002.

[DMvS97]   E. V. Dubrova, J. C. Muzio, and B. von Stengel. Finding Composition Trees for Multiple-valued Functions. In *Proceedings of the 27th International Symposium on Multiple-Valued Logic (ISMVL '97)*, pages 19–26, 1997.

[DP60]   M. Davis and H. Putnam. A Computing Procedure for Quantification Theory. In *Journal of ACM*, volume 7, pages 201–214, 1960.

[EB05]   N. Eén and A. Biere. Effective Preprocessing in SAT Through Variable and Clause Elimination. In *SAT*, 2005.

[EC81]   E. A. Emerson and E. M. Clarke. Characterizing Correctness Properties of Parallel Programs as Fixpoints. In *Proceedings of the 7th International Colloquium on Automata, Languages, and Programming*, number 85 in Proceedings of IEEE Symposium on Logic in Computer Science. Springer-Verlag, 1981.

[EL85]   E. A. Emerson and C. L. Lei. Modalities for Model Checking: Branching Time Strikes Back. In *Proceedings of ACM Symposium on Principles of Programming Languages*, pages 84–96, 1985.

[EL86]   E. A. Emerson and C. L. Lei. Efficient Model Checking in Fragments of the Propositional Mu-Calculus(Extended Abstract). In *Proceedings of IEEE Symposium on Logic in Computer Science*, pages 267–278, 1986.

[Eme90]   E. A. Emerson. Temporal and Modal Logic. In J. van Leeuwen, editor, *Formal Models and Semantics*, volume B of *Handbook of Theoretical Computer Science*, pages 996–1072. Elsevier Science, 1990.

[FAA01]   Karem A. Sakallah Fadi A. Aloul, Igor L. Markov. MINCE: A static global variable-ordering for SAT and BDD. In *Proceedings of International Workshop on Logic Synthesis*, June 2001.

[FDK98]   F. Fallah, S. Devadas, and K. Keutzer. Functional Vector Generation for HDL Models Using Linear Programming and 3-Satifiability. In *Proceedings of the Design Automation Conference (DAC)*, 1998.

[FDTM95] J. Freeman, R. Duerden, C. Taylor, and M. Miller. The 68060 microprocessor functional design and verification methodology. In *On-Chip Systems Design Conference*, pages 10.1–10.14, 1995.

[FKL97] H.D. Foster, A.C. Krolnik, and D.J. Lacey. *Assertion-Based Design*. Kluwer Academic Publishers, 1997.

[FOH93] H. Fujii, G. Ootomo, and C. Hori. Interleaving Based Variable Ordering Methods for Ordered Binary Decision Diagrams. In *Proceedings of International Conference on Computer-Aided Design (ICCAD)*, pages 38–41, November 1993.

[FORS01] J.-C. Filliatre, S. Owre, H. Rue, and N. Shankar. ICS: Integrated Canonizer and Solver. In *Proceedings of the Computer Aided Verification Conference (CAV)*, 2001.

[FS83] H. Fujiwara and T. Shimono. On the Acceleration of Test Generation Algorithms. In *IEEE Transactions on Computers*, volume C-32(12), December 1983.

[FS90] Steven J. Friedman and Kenneth J. Supowit. Finding the optimal variable ordering for binary decision diagrams. In *IEEE Transactions on Computers*, volume 39(5), pages 710–713, May 1990.

[FTKC91] M. Fujita, Y. Tamiya, Y. Kukimoto, and K.-C. Chen. Application of Boolean Unification to Combinational Logic Synthesis. In *Proceedings of International Conference on Computer-Aided Design (ICCAD)*, pages 510–513, 1991.

[FYBSV93] E. Felt, G. York, R. K. Brayton, and A. L. Sangiovanni-Vincentelli. Dynamic variable reordering for BDD minimization. In *Proceedings of European Conference on Design Automation (EDAC)*, pages 130–135, 1993.

[GAG+02] M.K. Ganai, P. Ashar, A. Gupta, L. Zhang, and S. Malik. Combining Strengths of Circuit-based and CNF-based Algorithms for a High-performance SAT Solver. In *Proceedings of the Design Automation Conference (DAC)*, pages 747–750, 2002.

[Gas79] J. Gaschnig. *Performance Measurement and Analysis of Certain Search Algorithms*. PhD thesis, Carnegie Mellon University, Department of Computer Science, May 1979.

[GF01] E. Gizdarski and H. Fujiwara. SPIRIT: A Highly Robust Combinational Test Generation Algorithm. In *Proceedings of IEEE VLSI Test Symposium (VTS)*, pages 346–351, April 2001.

[GG05] S.V. Gheorghita and R. Grigore. Constructing Checkers from PSL Properties. In *Proceedings of the 15th International Conference on Control Systems and Computer Science*, 2005.

[GGA04] M.K. Ganai, A. Gupta, and P. Ashar. Efficient SAT-based Unbounded Symbolic Model Checking Using Circuit Cofactoring. In *Proceedings of International Conference on Computer-Aided Design (ICCAD)*, pages 510–517, November 2004.

[GGYA03]   A. Gupta, M. Ganai, Z. Yang, and P. Ashar. Iterative Abstraction using SAT-based BMC with Proof Analysis. In *Proceedings of International Conference on Computer-Aided Design (ICCAD)*, pages 416–423, November 2003.

[GHS03]   M. Gordon, J. Hurd, and K. Slind. Executing the Formal Semantics of the Accellera Property Specification Language by Mechanised Theorem Proving. In *Proceedings of the 12th Conference on Correct Hardware Design and Verification Methods*, volume 2860 of *Lecture Notes in Computer Science*, pages 200–215. Springer-Verlag, 2003.

[GJ79]   M. R. Garey and D. S. Johnson. *Computers and Intractability*. W. H. Freeman and Co., 1979.

[GJC94]   M. Gyssens, P. Jeavons, and D. Cohen. Decomposing Constraint Satisfaction Problems Using Database Techniques. In *Artificial Intelligence*, pages 57–89, 1994.

[GL02]   D. Giannakopoulou and F. Lerda. From States to Transitions: Improving Translation of LTL Formula to B uchi Automata. In D. Peled and M.Y. Vardi, editors, *International Conference on Formal Techniques for Networked and Distributed Systems (FORTE'02)*, volume 2529 of *Lecture Notes in Computer Science*, pages 308–326. Springer-Verlag, 2002.

[GLS99]   G. Gottlob, N. Leone, and F. Scarcello. A Comparison of Structural CSP Decomposition Methods. In *IJCAI*, 1999.

[GN02]   E. Goldberg and Y. Novikov. BerkMin: a fast and robust SAT-solver. In *Proceedings of Design Automation and Test in Europe (DATE)*, pages 142–149, 2002.

[GO01]   P. Gastin and D. Oddoux. Fast LTL to Büchi Automata Translation. In G. Berry, H. Comon, and A. Finkel, editors, *Proceedings of the Computer Aided Verification Conference (CAV)*, volume 2012 of *Lecture Notes in Computer Science*, pages 53–65. Springer-Verlag, 2001.

[Goe81]   P. Goel. An Implicit Algorithm to Generate Tests for Combinational Logic Circuits. In *IEEE Transactions on Computers*, volume C-30(3), March 1981.

[Goe02]   R. Goering. Next-generation Verilog rises to higher abstraction levels. In *EE Times*, March 2002.

[Gor88]   M. Gordon. HOL: A Proof Generating System for Higher-order Logic. In G. Birwistle and P. A. Subrahmanyam, editors, *VLSI Specification, Verification and Synthesis*, pages 73–127. Academic Press, Boston, 1988.

[GSK98]   C.P. Gomes, B. Selman, and H. Kautz. Boosting Combinatorial Search Through Randomization. In *Proceedings of the National Conference on Artificial Intelligence*, 1998.

[Hav01]   J. Havlicek. Notes on blake canonical form based factorization. Technical report, Motorola, Inc., 2001. Unpublished report.

[HC00]   C.-Y. Huang and K.-T. Cheng. Assertion Checking by Combined Word-level ATPG and Modular Arithmetic Constraint-solving Techniques. In *Proceedings of the Design Automation Conference (DAC)*, pages 118–123, 2000.

[HFEM03]   J. Havlicek, D. Fisman, C. Eisner, and E. Marschner.   Mapping SVA to PSL.   In *Accellera*, October 2003.   Available at the URL: http://www.eda.org/vfv/docs/mapping.pdf.

[HKB96]   R. Hojati, S. C. Krishnan, and R. K. Brayton. Early Quantification and Partitioned Transition Relations. In *Proceedings of International Conference on Computer Design (ICCD)*, pages 12–19, October 1996.

[HMPS94]   G. D. Hachtel, E. Machii, A. Pardo, and F. Somenzi. Symbolic Algorithms to Calculate Steady-State Probabilities of a Finite State Machine. In *The European Design and Test Conference*, pages 214–218, 1994.

[Hoo94]   J.N. Hooker. Logic-based Methods for Optimization. In *ORSA CSTS Newsletter*, volume 15(2), pages 4–11, 1994.

[HvM04]   M. Heule and H. van Maare. Aligning CNF- and Equivalence-reasoning. In *SAT*, 2004.

[HW05]   J. Havlicek and Y. Wolfsthal. PSL and SVA: Two Standard Assertion Languages Addressing Complementary Engineering Needs. In *DVCon2005*, 2005.

[IPC05]   M.K. Iyer, G. Parthasarathy, and K.-T. Cheng. Efficient Conflict Based Learning in a Constraint Solver for RTL Circuits. In *Proceedings of Design Automation and Test in Europe (DATE)*, pages 666–671, 2005.

[Iye03]   M.A. Iyer. RACE: A Word-level ATPG-based Constraints Solver System for Smart Random Simulation. In *Proceedings of International Test Conference (ITC)*, pages 299–308, 2003.

[JD01]   P. Johannsen and R. Drechsler. Formal Verification on the RT Level Computing One-to-one Design Abstractions by Signal Width Reduction. In *In Proceedings of IFIP International Conference on Very Large Scale Integration(IFIP VLSI-SOC 2001)*, 2001.

[JM94]   J. Jaffar and M.J. Maher. Constraint Logic Programming: A Survey. In *Journal of Logic Programming*, volume 19/20, pages 503–581, 1994.

[JS97]   R. Bayardo Jr. and R. Schrag. Using CSP Look-Back Techniques to Solve Real-World SAT Instances. In *Proceedings of the 14th National Conference on Artificial Intelligence and 9th Innovative Applications of Artificial Intelligence*, pages 203–208. AAAI press, July 1997.

[KBS93]   R. Krieger, B. Becker, and R. Sinkovic. A BDD-based Algorithm for Computation of Exact Fault Detection Probabilities. In *International Symposium on Fault-Tolerant Computing*, pages 186–195, 1993.

[KM94]   D. Koller and N. Megiddo. Constructing small sample spaces satisfying given constraints. In *SIAM Journal on Discrete Mathematics*, pages 260–274, 1994.

[KM96]   M. Kaufmann and J. S. Moore. ACL2: An Industrial Strength Version of Nqthm.
            In *Proceedings of the 11th Annual Conference on Computer Assurance, IEEE
            Computer Society press*, pages 23–34, June 1996.

[KP94]   W. Kunz and D. Pradhan. Recursive learning: A precise implication procedure
            and its application to test verification and optimization. In *IEEE Transactions on
            Computer-Aided Design of Integrated Circuits and Systems*, September 1994.

[KS00]   J.H. Kukula and T.R. Shiple. Building Circuits From Relations. In *Proceedings
            of the Computer Aided Verification Conference (CAV)*, 2000.

[Kur87]   R. P. Kurshan. Reducibility in Analysis of Coordination. In *Discrete Event Sys-
            tems: Models and Applications*, volume 103 of *LNCS*, pages 19–39. Springer-
            Verlag, 1987.

[Kur93]   R. P. Kurshan. *Automata-Theoretic Verification of Coordinating Processes*.
            Princeton University Press, 1993.

[Kur94]   R. P. Kurshan. Computer-Aided Verification of Coordinating Processes: The
            Automata-Theoretic Approach. In *Princeton University Press*, 1994.

[KV99]   O. Kupferman and M.Y. Vardi. Model Checking of Safety Properties. In
            Proceedings of the Computer Aided Verification Conference (CAV), volume
            1633 of *Lecture Notes in Computer Science*, pages 172–183. Springer-Verlag,
            1999.

[KV01]   O. Kupferman and M.Y. Vardi. On Clopen Specifications. In *Proceedings of the
            8th International Conference on Logic for Programming, Artificial Intelligence,
            and Reasoning*, Lecture Notes in Computer Science. Springer-Verlag, 2001.

[Lee59]   C. Y. Lee. Representation of Switching Circuits by Binary-decision Programs.
            In *Bell System Technical Journal*, volume 38, No. 4, pages 985–999, July 1959.

[LK03]   C. Liu and A. Kuehlmann. CAMA: A Multi-Valued Satisfiability Solver. In
            Proceedings of International Conference on Computer-Aided Design (ICCAD),
            pages 326–333, November 2003.

[LL92]   H.T. Lau and C.-S. Lim. On the OBDD Representation of General Boolean
            Functions. In *IEEE Transactions on Computers*, volume 41(6), pages 661–663,
            1992.

[Löw08]   L. Löwenheim. Über das Auflösungsproblem im logischen Klassenkalkül. In
            Sitzungsber. Berl. Math. Gessel. 7, pages 89–94, 1908.

[LS81]   R.J. Lipton and R. Sedgewick. Lower Bounds for VLSI. In *Proceedings of
            ACM Symposium on the Theory of Computing*, pages 300–307, May 1981.

[MA03]   K.L. McMillan and N. Amla. Automatic Abstraction without Counterexam-
            ples. In *Proceedings of Tools for Algorithms for Construction and Analysis of
            Systems (TACAS)*, 2003.

[MBC$^+$00] D. MacMillen, M. Butts, R. Camposano, D. Hill, and T.W. Williams. An
            Industrial View of Electronic Design Automation. In *IEEE Transactions on*

Computer-Aided Design of Integrated Circuits and Systems, volume 19(12), pages 1428–1448, December 2000.

[McM93]   Kenneth L. McMillan. *Symbolic Model Checking*. Kluwer Academic Publishers, 1993.

[McM96]   K. L. McMillan. A Conjunctively Decomposed Boolean Representation for Symbolic Model Checking. In *Proceedings of the Computer Aided Verification Conference (CAV)*, pages 13–25, 1996.

[McM02]   K. McMillan. Applying SAT methods in unbounded symbolic model checking. In *Proceedings of the Computer Aided Verification Conference (CAV)*, pages 250–264, 2002.

[McM03]   K. L. McMillan. Interpolation and SAT-based Model Checking. In *Proceedings of the Computer Aided Verification Conference (CAV)*, July 2003.

[McN66]   R. McNaughton. Testing and Generating Infinite Sequences by a Finite Automaton. In *Information and Control*, volume 9, pages 521–530, 1966.

[MDA96]   C-J. H. Segar M. D. Aagaard, R. B. Jones. Formal Verification Using Parametric Representations of Boolean Constraints. In *Proceedings of the Design Automation Conference (DAC)*, 1996.

[MFM04]   Y.S. Mahajan, Z. Fu, and S. Malik. Zchaff 2004: An efficient SAT solver. In *SAT*, Lecture Notes in Computer Science. Springer-Verlag, 2004.

[Mil93]   D.M Miller. Multiple-valued Logic Design Tools. In *Proceedings of the 23rd International Symposium on Multiple-Valued Logic*, pages 2–11, 1993.

[MLK98]   J.S. Moore, T. Lynch, and M. Kaufmann. A Mechanically Checked Proof of the Correctness of the Kernel of the AMD5K86 Floating Point Division Algorithm. In *IEEE Transactions on Computers*, volume 47, pages 913–926, 1998.

[MMZ+01]   M.W. Moskewicz, C.F. Madigan, Y. Zhao, L. Zhang, and S. Malik. Chaff: Engineering an Efficient SAT Solver. In *Proceedings of the Design Automation Conference (DAC)*, July 2001.

[MN89]   U. Martin and T. Nipkow. Boolean Unification - the Story So Far. In *Journal of Symbolic Computation*, volume 7, pages 275–293, 1989.

[MS99]   J.P. Marques-Silva. The Impact of Branching Heuristics in Propositional Satisfiability Algorithms. In *Proceedings of the 9th Portuguese Conference on Artificial Intelligence*, September 1999.

[MSP01]   A. Mishchenko, B. Steinbach, and M. Perkowski. An Algorithm for Bi-Decomposition of Logic Functions. In *Proceedings of the Design Automation Conference (DAC)*, pages 103–108, 2001.

[MSS96]   J. P. Marques-Silva and K. Sakallah. GRASP–A New Search Algorithm For Satisfiability. In *Proceedings of International Conference on Computer-Aided Design (ICCAD)*, Santa Clara, CA, November 1996.

[MY60]   R. McNaughton and H. Yamada. Regular Expression and State Graphs for Automata. In *IRA Transaction on Electronic Computers*, volume EC-9(1), pages 39–47, 1960.

[NO79]   G. Nelson and D. Oppen. Simplification by Cooperating Decision Procedures. In *ACM Transactions on Programming Languages and Systems*, volume 1(2), pages 245–257, 1979.

[ORSvH95]   S. Owre, J. Rushby, N. Shankar, and F. von Henke. Formal Verification for Fault-tolerant Architectures: Prolegomena to the Design of PVS. In *IEEE Transaction on Software Engineering*, volume 21(2), pages 107–125, February 1995.

[Par70]   D. Park. Fixpoint Induction and Proof of Program Semantics. In B. Meltzer and D. Michie, editors, *Machine Intelligence*, volume 5, pages 59–78, Edinburgh, 1970. Edinburgh University Press.

[PICB05]   G. Parthasarathy, M.K. Iyer, K.-T. Cheng, and F. Brewer. Structural Search for RTL with Predicate Learning. In *Proceedings of the Design Automation Conference (DAC)*, pages 451–456, 2005.

[PICW04]   G. Parthasarathy, M.K. Iyer, K.-T. Cheng, and L.-C. Wang. An Efficient Finite Domain Constraint Solver for Circuits. In *Proceedings of the Design Automation Conference (DAC)*, pages 212–217, 2004.

[Pix92]   C. Pixley. A Theory and Implementation of Sequential Hardware Equivalence. In *IEEE Transactions on Computer-Aided Design of Integrated Circuits and Systems*, volume 11, pages 1469–1494, December 1992.

[Pix99]   C. Pixley. Integrating Model Checking Into the Semiconductor Design Flow. In *Journal of Electronic Systems*, pages 67–73, March 1999.

[PS94]   S. Panda and F. Somenzi. Symmetry Detection and Dynamic Variable Ordering of Decision Diagrams. In *Proceedings of International Conference on Computer-Aided Design (ICCAD)*, 1994.

[Pug92]   W. Pugh. The Omega Test: a Fast and Practical Integer Programming Algorithm for Dependence Analysis. In *Communication of the ACM*, August 1992.

[RAP+95]   R. K. Ranjan, A. Aziz, B. Plessier, C. Pixley, and R. K. Brayton. Efficient BDD Algorithms for FSM Synthesis and Verification. In *Proceedings of International Workshop on Logic Synthesis*, May 1995.

[RBS67]   J.P. Roth, W.G. Bouricius, and P.R. Schneider. Programmed Algorithms to Compute Tests to Detect and Distinguish Between Failures. In *IEEE Transaction on Electronic Computers*, volume EC-16(10), October 1967.

[Rot66]   J.P. Roth. Diagnosis of Automata Failures: A Calculus and Method. In *IBM Journal of Research and Development*, volume 10(4), July 1966.

[RS86]   N. Robertson and P. D. Seymour. Graph minors. II. Algorithmic aspects of tree-width. In *Journal of Algorithms*, pages 7:309–322, 1986.

[Rud74]   S. Rudeanu. *Boolean Functions and Equations*. North-Holland, 1974.

[Rud93]   R. Rudell. Dynamic Variable Ordering for Binary Decision Diagrams. In *Proceedings of International Conference on Computer-Aided Design (ICCAD)*, pages 42–47, November 1993.

[Saf88]   S. Safra. On Complexity of ω-automata. In *Proceedings of 29th IEEE Symposium on Foundations of Computer Science*, pages 319–327, 1988.

[SB00]   F. Somenzi and R. Bloem. Efficient Büchi Automata from LTL Formulae. In E.A. Emerson and A.P. Sistla, editors, *Proceedings of the Computer Aided Verification Conference (CAV)*, volume 1633 of *Lecture Notes in Computer Science*, pages 257–263. Springer-Verlag, December 2000.

[Sch86]   A. Schrijver. *Theory of Linear and Integer Programming*. Wiley, 1986.

[Set al.]   F. Somenzi and *et al*. CUDD: CU Decision Diagram Package. Technical report, University of Colorado. ftp://vlsi.colorado.edu/pub/.

[SGZ+98]   Khurram Sajid, Anuj Goel, Hai Zhou, Adnan Aziz, Suzanne Barber, and Vigyan Singhal. Symbolic Procedures for a Theory of Equality with Uninterpreted Functions. Submitted to CAV 1998, 1998.

[Sho84]   R. Shostak. Deciding Combinations of Theories. In *Journal of the ACM*, volume 31(1), pages 1–12, 1984.

[SHSVB94]   T. R. Shiple, R. Hojati, A. L. Sangiovanni-Vincentelli, and R. K. Brayton. Heuristic Minimization of BDDs Using Don't Cares. In *Proceedings of the Design Automation Conference (DAC)*, pages 225–231, San Diego, CA, June 1994.

[SI98]   CA. Synopsys Inc., Mountain View. Multimillion-gate ASIC verification. 1998.

[Sie84]   J. H. Siekmann. Universal Unification. In *7th International Conference on Automated Deduction*, Lecture Notes in Computer Science, pages 1–42, 1984.

[SKMB90]   A. Srinivasan, T. Kam, S. Malik, and R.E. Brayton. Algorithms for Discrete Function Manipulation. In *Proceedings of International Conference on Computer-Aided Design (ICCAD)*, pages 92–95, 1990.

[SLB03]   S. Seshia, S. Lahiri, and R. Bryant. A Hybrid SAT-Based Decision Procedure for Separation Logic with Uninterpreted Functions. In *Proceedings of the Design Automation Conference (DAC)*, pages 425–430, 2003.

[SP04]   S. Subbarayan and D.K. Pradhan. NiVER: Non Increasing Variable Elimination Resolution for Preprocessing SAT Instances. In *SAT*, 2004.

[SS77]   R.M. Stallman and G.J. Sussman. Forward Reasoning and Dependency-Directed Backtracking in a System for Computer-Aided Circuit Analysis. In *Artificial Intelligence*, volume 9, pages 135–196, 1977.

[SS05]   H.M. Sheini and K.A. Sakallah. Pueblo: A Modern Pseudo-Boolean SAT Solver. In *Proceedings of Design Automation and Test in Europe (DATE)*, pages 684–685, 2005.

[Sta92] G. Stalmarck. A System for Determining Propositional Logic Theorems by Applying Values and Rules to Triplets That Are Generated From Boolean Formula. 1992. Swedish Patent No. 467,076 (approved 1992), U.S. Patent No. 5,276,897 (approved 1994), European Patent No. 0403 454 (approved 1995).

[TBK90] H. Touati, R. K. Brayton, and R. P. Kurshan. Checking Language Containment using BDDs. In *Proceedings of International Workshop on Formal Methods in VLSI Design*, Miami, FL, January 1990.

[TGH97] P. Tafertshofer, A. Ganz, and M. Hentfling. A SAT-Based Implication Engine for Efficient ATPG, Equivalence Checking, Optimization of Netlists. In *Proceedings of International Conference on Computer-Aided Design (ICCAD)*, pages 648–655, November 1997.

[Thi02] X. Thirious. Simple and Efficient Translation from LTL Formulas to Büchi Automata. In *Workshop in Formal Methods for Industrial Critical Systems (FMICS'02)*, volume 66(2) of *ENTCS*, pages 308–326. Springer-Verlag, July 2002.

[Tho68] K. Thompson. Regular Expression Search Algorithms. In *Communication of ACM*, volume 11(6), pages 419–422, 1968.

[TY84] R. E. Tarjan and M. Yannakakis. Simple linear-time algorithms to test chordality of graphs, test acyclicity of hypergraphs and selectively reduce acyclic hypergraphs. In *SIAM J. Comput.*, volume **13**(3), pages 566–579, 1984.

[Vel04] M.N. Velev. Using Automatic Case Splits and Efficient CNF Translation to Guide a SAT-solver When Formally Verifying Out-of-order Processors. In *8th International Symposium on Artificial Intelligence and Mathematics*, 2004.

[W. 87] W. Büttner and H. Simonis. Embedding Boolean Expressions into Logic Programming. In *Journal of Symbolic Computation*, volume 4, pages 191–205, 1987.

[W. 88] W. Büttner. Unification in Finite Algebras Is Unitary. In *Proceedings of CADE-9*, volume 310 of *Lecture Notes in Computer Science*, pages 368–377, 1988.

[Wil76] H.P. Williams. Fourier-Motzkin Elimination Extension to Integer Programming Problems. In *Journal of Combinatorial Theory*, volume 21, 1976.

[Wol83] P. Wolper. Temporal logic can be more expressive. In *Information and Computation (formerly Information and Control)*, volume 56, 1983.

[WVS83] P. Wolper, M.Y. Vardi, and A.P. Sistla. Reasoning about Infinite Computation Paths. In *Proceedings of 24th IEEE Symposium on Foundations of Computer Science*, pages 185–194, 1983.

[YA00] J. Yuan and A. Aziz. Random Vector Generation Using Event Probabilities. In *Technical report*, 2000.

[YAA02] J. Yuan, A. Aziz, and K. Albin. Enhancing Simulation Coverage Through Guided Vector Generation. In *Technical report*, 2002.

[YAAP03]   J. Yuan, K. Albin, A. Aziz, and C. Pixley. Constraint Synthesis for Environment Modeling in Functional Verification. In *Proceedings of the Design Automation Conference (DAC)*, June 2003.

[YAAP04]   J. Yuan, K. Albin, A. Aziz, and C. Pixley. Simplifying Boolean Constraint Solving By Extracting Hold-Constraints. In *IEEE Transactions on Computer-Aided Design of Integrated Circuits and Systems*, March 2004.

[YAS00]   P. Yalagandula, A. Aziz, and V. Singhal. Automatic Lighthouse Generation for Directed State Space Search. In *Proceedings of Design Automation and Test in Europe (DATE)*, March 2000.

[YD98]   C. H. Yang and D. L. Dill. Validation with Guided Search of the State Space. In *Proceedings of the Design Automation Conference (DAC)*, pages 599–604, 1998.

[YH01]   J. Yuan and J. Havlicek. Prime cover based constraint factorization in simgen. Technical report, Motorola, Inc., 2001. Unpublished report.

[YKAP02]   J. Yuan, A. Aziz K. Albin, and C. Pixley. Simplifying Boolean Constraint Solving For Random Simulation-Vector Generation. In *Proceedings of International Conference on Computer-Aided Design (ICCAD)*, pages 123–127, 2002.

[YPAA03]   J. Yuan, C. Pixley, A. Aziz, and K. Albin. A Framework for Constrained Functional Verification. In *Proceedings of International Conference on Computer-Aided Design (ICCAD)*, pages 142–145, November 2003.

[YSAA97]   J. Yuan, J. Shen, J. A. Abraham, and A. Aziz. On Combining Formal and Informal Verification. In *Proceedings of the Computer Aided Verification Conference (CAV)*, Lecture Notes in Computer Science, pages 376–387. Springer-Verlag, June 1997.

[YSBO99]   B. Yang, R. Simmons, R.R. Bryant, and D.R. O'Hallaron. Optimizing Symbolic Model Checking for Constraint-rich Models. In *Proceedings of the Computer Aided Verification Conference (CAV)*, pages 328–340, 1999.

[YSP$^+$99]   J. Yuan, K. Shultz, C. Pixley, H. Miller, and A. Aziz. Modeling Design Constraints and Biasing in Simulation Using BDDs. In *Proceedings of International Conference on Computer-Aided Design (ICCAD)*, pages 584–589, 1999.

[YSPM01]   J. Yuan, K. Shultz, C. Pixley, and H. Miller. Method and Apparatus for Integrated Circuit Design Verification. granted in November, 2001. US Patent No. 6321186.

[Yu92]   Y. Yu. Automated Correctness Proofs of Machine Code Programs for a Commercial Microprocessor. In *Automated Deduction – CADE-11, Lecture Notes in Computer Science*, pages 416–430. Springer-Verlag, 1992.

[ZCR01]   Z. Zeng, M. Ciesielski, and B. Rouzeyre. Functional Test Generation Using Constraint Logic Programming. In *In Proceedings of IFIP International Conference on Very Large Scale Integration (IFIP VLSI-SOC 2001)*, 2001.

[Zha97]   H. Zhang. SATO: An efficient propositional prover. In *Proceedings of the International Conference on Automated Deduction*, pages 272–275, July 1997.

[ZKC01]   Z. Zeng, P. Kalla, and M. Ciesielski. LPSAT: A Unified Approach to RTL Satisfiability. In *Proceedings of Design Automation and Test in Europe (DATE)*, 2001.

[ZM03a]   L. Zhang and S. Malik. Extracting Small Unsatisfiable Cores from Unsatisfiable Boolean Formula. In *SAT*, 2003.

[ZM03b]   L. Zhang and S. Malik. Validating SAT Solvers Using an Independent Resolution Based-Checker: Practical Implementations and Other Applications. In *Proceedings of Design Automation and Test in Europe (DATE)*, 2003.

[ZMMM01]   L. Zhang, C.F. Madigan, M. W. Moskewicz, and S. Malik. Efficient Conflict Driven Learning in a Boolean Satisfiability Solver. In *Proceedings of International Conference on Computer-Aided Design (ICCAD)*, November 2001.

Index